# The Road to Quality

D. M. Lascelles
B. G. Dale

IFS Ltd., UK
1993

P0 3726

**British Library Cataloguing-in-Publication Data**

A catalogue record for this book is available from the British Library.

ISBN 1-85423-103-0



Photo-typeset by A J Latham Ltd., Houghton Regis, Dunstable, UK.

Printed by Information Press Ltd., Oxford, UK.

# Contents

# Preface

Total Quality Management is concerned with managing change: changing customer requirements; changing competitive forces; and changing organisational structures, processes, systems, behaviour and attitudes. At its heart is people: how they think and act (culture); how they are managed (organisation); and how they work (processes). Therefore, the central themes of this book are people and change.

This book is not about the theory of organisational change, nor is it about the theory of Total Quality Management. This is a book about the realities of initiating and sustaining Total Quality as a business improvement strategy; a strategy with the overall goal of complete customer satisfaction. The contents of the book are based on eleven or so years of research, consultancy and practical experience of TQM.

Although each chapter is self-contained and focuses on a specific theme, the book is structured in an ordered sequence. In Chapter 1 we set the scene by defining quality, and discuss how the activities of quality management are shaped by the three parameters of customer perception, competition and business performance. The modern concept of TQM has evolved from traditional methods of quality management to one of an organisation-wide set of cultural values dedicated to releasing the talents of everyone to actively increase customer satisfaction and corporate competitiveness. We have introduced the term 'Total Quality Improvement' to describe the incremental change strategy for achieving a TQM culture.

The leading Japanese companies have developed such a culture, in which continuous improvement is a way of organisational life. Therefore, we present in Chapter 2 the findings of four guided study tours of European executives to Japan to illustrate how the Japanese achieve this in practice. World-class competitiveness is not a product of national culture. The Japanese perform no special cultural magic, but adhere to a simple formula: create a clear strategic direction, apply it systematically throughout the organisation, stick to it, continuously improve, constantly analyse data, and educate and empower people.

Chapter 3 examines the forces which motivate the kind of organisational impetus required to trigger the process of Total Quality Improvement, and traces their effects on five different companies. What emerges is the fact that the most important forces for motivating and then sustaining change are demanding customers and the chief executive; two key change agents we refer to throughout the rest of the book.

Having started down the quality improvement road there is no sure-fire recipe for developing and sustaining the process. All too often the quick-fix predominates. Many managers seem to yearn for a package they can plug-in, switch-on and leave to convert desire into world-class quality. The pros and cons of the commonest approaches to starting quality improvement are discussed in Chapter 4. We also describe how three companies successfully introduced quality improvement. Their secret? A clear strategic direction, and committed and involved management leadership.

Leadership is about coping with change. Leadership is, therefore, the subject of Chapter 5. The nature of leadership and its role in Total Quality Improvement is examined. Without effective leadership, TQI initiatives do not succeed. A Total Quality culture can evolve by giving as many people as possible leadership opportunities – this is called 'empowerment'. We also describe the actions of three chief executives who demonstrated leadership through Total Quality, and include a road map for would-be Total Quality leaders.

At some point in the quality improvement process, usually sooner rather than later, most organisations want to improve the performance of their suppliers. Working with suppliers to resolve problems and forming long-term business partnerships with them should be a featured goal of every Total Quality Improvement strategy. But, 'caveat emptor', most supply problems can be caused by the buyer, therefore, it is necessary to first put one's own house in order before attempting to do the same for one's suppliers. Chapter 6 covers the issues of supply-chain management and what is involved in preparing for and implementing a supplier development initiative.

Total Quality Improvement is a long-term commitment, and it can take up to ten years to achieve a significant cultural change in which the principles of Total Quality Management are put firmly in place. In Chapter 7 we examine the difficulties and barriers to sustained quality improvement. A case study of failure accounts for approximately half of the chapter. We have included this case study to illustrate Deming's point that quality improvement – and hence survival – is not compulsory.

Awareness, communication and the challenge of technology transfer are critical factors in stimulating the changes in attitudes and behaviour on which a Total Quality culture depends. National initiatives to stimulate 'the pursuit of excellence' have had a significant impact on raising the profile of quality in the public consciousness. In Chapter 8 we record the effects of one such initiative – the UK National Quality Campaign – and discuss its lessons together with some of the issues involved in moving from awareness to sustained action. We review the scene in the rest of Europe and the USA, particularly with regard to the development of two models intended to underpin the process of TQM-related technology transfer: The Malcolm Baldrige National Quality Award and The European Quality Award for Companies.

In our experience there is nothing mystical about Total Quality Improvement. To be successful it requires just a few simple elements: top management commitment and participation; clarity of vision; careful planning; company-wide training and education linked to specific skill needs and improvement objectives; an integrated deployment structure to ensure continuity of the process; and patience. This is not a menu; all the elements must be present. In Chapter 9 we outline the eight steps that make up a strategic framework for Total Quality Improvement.

The extent to which companies have adopted the principles and values of Total Quality has been extremely variable over the last decade. Many companies have no clear idea of the progress they have made or how far they still have to go. In an attempt to remove some of the confusion, we present in Chapter 10 six levels of Total Quality Improvement adoption, or lack of it. You might like to see where your organisation lines up.

Perhaps it is unlikely that a chief executive officer will read the book from cover-to-cover. They will, however, benefit from using the text as a reference to provide guidance on specific aspects of quality improvement. Executives will find the summary of key points section of particular value for easy reference.

The book will appeal most to those members of the executive management team who have responsibility for strategic quality planning, and introducing and developing quality improvement initiatives. Undergraduates, post-graduates and those studying for professional examinations which involve Total Quality should also find the book a valuable source of reference, as will the increasing numbers of the academic community who have a teaching and research interest in the subject.

So either read the book in chapter sequence or dip into it as you please – it's your book, the choice is yours.

**David Lascelles**
**Barrie Dale**

# Acknowledgements

We wish to thank our UMIST colleague, Dr. Ruth Boaden, who read the complete manuscript and made many constructive comments on how it could be improved.

We also express our thanks to: Mike Collins, Ian Ferguson, Bernard Payne, Roy Peacock and David Pentecost who also provided many invaluable hints on improving the manuscript.

Finally, but not least, our grateful thanks to Mrs. Irene Cunningham who typed the manuscript.

# Summary of Key Points

## Chapter 1 – What is Quality?

- Quality means different things to different people and there are many definitions and interpretations of the word.
- Quality is often confused with grade.
- Quality is what gives complete customer satisfaction.
- Quality is a key business strategy of superior performing organisations.
- Three parameters impact on the set of activities called quality management – customer perception, competition and business performance.
- Inspection, quality control and quality assurance are phases out of which TQM has evolved.
- TQM involves the application of quality management principles to all aspects of a business.
- TQM requires a strategic commitment by the chief executive officer and members of the senior management team, meaning a fundamental change in behaviour, attitudes, values, culture, organisational structure and work practices.
- The key features of TQM are:
  - Everyone accepting ownership for the continuous improvement of their processes.
  - Recognising that everyone for whom a task is carried out is a customer and ensuring that these customers are satisfied.
  - Measurement and benchmarking of performance is continuous.
  - Involvement of people in the business.
  - Development of people.
  - Teamworking.
  - No departmental barriers.
  - Standardisation and simplification of work practises.
- TQM is not a programme – it is a process which is continuous.
- Kaizen is the epitome of continuous improvement.
- Total Quality Improvement (TQI) is a business strategy to achieve the goal of complete customer satisfaction. TQI occurs when everyone and every function is involved and committed to

continuous quality improvement with customers and suppliers involved in the process.

- TQI has three dimensions: improvement, standardisation and the strategic dynamics of TQM.
- The strategic dynamics of TQM are:
  - Customer care.
  - Competitors.
  - Leadership.
  - Suppliers.
  - People.
  - Money.

## Chapter 2 – Total Quality Commitment: The Japanese Approach

- Japanese companies are totally dedicated to satisfying the needs and requirements of their customers, and this effort is long-term and continuous.
- A variety of methods are used for listening to customers in order to pinpoint the technical gaps in relation to the competition, and identify product features which the customer finds attractive
- Quality, cost and delivery are the main objectives, and organisations strive to become the best in class in relation to these three criteria.
- Planning, feedback and decision-making on Total Quality Control (TQC) is long-term, often extending for at least ten years into the future.
- Policy deployment is a key TQC planning activity.
- Research and development is the key by which Japanese companies sustain their competitive edge. The aim is to bring an increasing number of attractive and high-value added products to the market in the shortest possible time.
- The facilitation of continuous improvement and promoting TQC is separated out from quality assurance activities.
- Quality assurance is the central core of TQC, and without effective quality assurance procedures, TQC is difficult.
- When problems occur, Japanese companies place considerable emphasis on finding out where and why things have gone wrong.
- The Japanese have created an organisational culture in which quality improvement is automatic, and a natural instinct.
- Japanese companies place considerable emphasis on ensuring

that operating data are visible on the factory shopfloor, believing that everyone benefits from an open communication system.

- Quality control circles and suggestion schemes are important mechanisms and motivators for involving everyone inside an organisation in continuous improvement.
- Japanese companies are totally committed to a long-term programme of TQC education and training for all employees, including senior management.
- Total Productive Maintenance is actively pursued to improve the quality and efficiency of machinery, and this supports TQC.
- Just-in-Time is considered by Japanese companies to be a key feature of TQC.

## Chapter 3 – The Motivation for Total Quality Improvement

- TQI is a form of organisational change.
- Organisational effectiveness depends on commonality of purpose and structure.
- Change catches organisations on the 'horns' of a dilemma between the desire for change and the desire to preserve continuity.
- Organisational change can never be spontaneous and there must be good reasons for it to happen.
- Demanding customers are the key external force or trigger for motivating quality improvement activity, and the chief executive officer is the key internal force.
- Simulating the conditions of a greenfield or restart opportunity can be a useful technique of change.
- There are six forces which are the main triggers of quality improvement: the chief executive officer, competition, changes in the cost structure of the market-place, demanding customers, a greenfield venture, and a restart situation.
- The forces of change are links in a chain. Competition acts as a catalyst setting off a chain-reaction which enhances quality awareness in the market, resulting in demanding customers and chief executives behaving as change agents.
- The motivators of quality improvement can be put into three categories:
  1. Catalysts of change – competition, costs.
  2. Change agents – demanding customers, the chief executive.
  3. Change opportunities – greenfield ventures, restart situations.

## Chapter 4 – How Companies Start Quality Improvement

- There is no single best way of starting a process of quality improvement which suits all organisations and cultures.
- A considerable amount of conflicting advice is given to organisations, and this can hamper the spread and development of quality improvement.
- The main approaches to getting started on quality improvement are: quality management tools and techniques, the teachings of the quality management gurus, the use of a management consultant, and quality system development and registration.
- Quality management tools and techniques fulfil a number of roles, and it is important that managers are fully aware of the main purpose and use of the tools and techniques they are considering applying.
- It is a mistake to single out one tool or technique for special attention.
- It is important that a planned approach is developed for the application of tools and techniques.
- The ways of approaching TQM suggested by Crosby, Deming, Feigenbaum and Juran are variations on a theme, the essential difference between them being the focus of their approach.
- The teachings of these four gurus can be characterised by the main focus of their approach:
  - Crosby: company-wide motivation.
  - Deming: statistical process control.
  - Feigenbaum: systems management.
  - Juran: project management.
- Organisations do not usually stick with the teachings of just one quality management expert.
- The use of management consultancy can be of assistance in getting the quality process under way, but it does not relieve the board of directors of its own responsibilities for TQM.
- Quality system registration to a recognised quality system standard is an essential building block of TQM, but it should never be perceived by an organisation as the pinnacle of its TQM efforts.
- The method of getting started on quality improvement is less important than management's commitment to continuous quality improvement, and the leadership it is prepared to demonstrate, this is the key determinant of long-term success.

## Chapter 5 – Leadership

- Leadership is about coping with change.
- People within an organisation expect their leaders to act as role-models and symbols, and to take their cue from their leaders' behaviour. Leaders are potent symbols.
- Leadership and management represent different but complementary activities. Leadership is inspirational, breaking new ground. Management is concerned with minimising risk and failure.
- To become effective leaders, managers must learn to perform the core tasks of:
  - Setting a direction.
  - Aligning people.
  - Motivating and inspiring people.
- Good leaders are usually in short supply. Companies should seek to develop their own by evolving a leadership culture in which people are given training and opportunities to lead and facilitate change.
- An increasing number of chief executive officers are seeing product and service quality as a key competitive factor in business performance.
- The chief executive officer has a crucial role in leading the quality improvement process, and must exercise strong leadership. This involves two key tasks: shaping organisational values and establishing a managerial infrastructure to actually bring about change. However, many chief executives are not fully committed to taking up this leadership role.
- There are a number of good role-models who can provide the inspiration for the chief executive officer wishing to lead the TQI initiative.
- The leadership criteria of the European Quality Award can also provide some useful pointers concerning what is required.

## Chapter 6 – Supply-Chain Management

- Supply-chain management is a key issue because:
  - The quality of purchased supplies is critical to the quality of a company's product or service.

  - The purchase of materials from suppliers and sub-contractors can account for a least 50% of manufacturing costs.
  - Technological and competitive pressures have resulted in a greater trend towards specialisation.
  - Those companies with the best suppliers, and which make the most effective use of their capabilities, will have a competitive advantage.
- There are definite trends towards the establishment of long-term business relationships between customers and suppliers. Many major purchasing organisations are committed to this change, because without it they cannot hope to stay ahead of the competition.
- To develop the relationship, considerable changes in behaviour and attitude are required in both customer and supplier organisations; it will take some time for these changes to become effective and permanent. Failure will occur when there is lack of commitment on one or both sides.
- The main barriers to supplier development are:
  - Poor communication and feedback.
  - Supplier complacency.
  - Misguided supplier improvement objectives.
  - Lack of customer credibility.
  - A misconception of purchasing power.
- Before starting supplier development, a purchasing organisation should give attention to internal issues such as:
  - The objectives of supplier development.
  - Developing a strategy and plan to achieve the objectives.
  - Deciding which vendors to involve.
- Supplier development starts with a review of those key aspects of the purchasing organisation's own operation which effect supplier performance. These aspects include:
  - Purchase specifications.
  - Communications.
  - Training.
  - Roles of different departments.
- The supplier development process includes:
  - Obtaining the commitment of suppliers.
  - Executive communication.
  - Supplier assessment.
  - Recognition of supplier performance.

## Chapter 7 – Difficulties and Barriers to Quality Improvement

- At the outset, many management teams fail to realise that TQI is a long-term commitment. It can take up to ten years to put some of the fundamental principles into place and create the right organisational culture. When they do recognise this, disillusionment can set in.
- Most organisations are likely to face difficulties in starting and sustaining quality improvement. These difficulties can be boiled down to:
  - Company culture.
  - Lack of senior management commitment.
  - Ineffective leadership.
- The main barriers to improvement are:
  - Management style: transactional (reacting to events) rather than transforming (shaping future events).
  - Fear of change.
  - Lack of the necessary skills to facilitate improvement.
  - Lack of information to support the improvement process.
- Senior management are the people with the most power to resist change. Managerial resistance to change has two root causes:
  - Change is perceived as a threat to and erosion of their authority.
  - A marked reluctance on the part of managers to examine what they actually do.
- A planned programme of training and education in quality-related skills for all employees, including senior management, is a prerequisite for the success of a quality improvement process.
- The skill needs of quality improvement leaders are often taken for granted.
- The effectiveness of the quality improvement process is dependent on the availability of relevant, reliable and objective information.
- If the normal agents of change are missing an organisation's management is likely to be complacent and inward looking, to pursue the interests of their particular department and functions, and to be in conflict over the key issues facing the organisation. They will also lack the data on which to base decisions. In such situations quality improvement will be slow in coming and difficult to sustain.

- People within organisations are usually more committed to improvement than top management realise, but lack the opportunity and resources.

## Chapter 8 – Quality Awareness, Communication and Challenges

- The UK's National Quality Campaign has increased her industry's awareness of the importance of product and service quality, and has helped to stimulate a positive reaction within business.
- Senior managers do read quality management publications, but are selective in their choice. They are primarily interested in material which will assist them to improve business performance.
- Industrialists have high expectations of the role of government in promoting quality management.
- An infrastructure for the transfer of TQM experiences and technology requires a partnership between government, industry, academic institutions, management consultants and professional institutions. The Union of Japanese Scientists and Engineers (JUSE) provides a well founded model for such an infrastructure.
- Facilitating technology transfer of quality-related information is not easy, and this difficulty is compounded by a variety of vested interests. Centres of Excellence, Networks, Partnerships, Quality Clubs and Q-Share can be important transfer media.
- Technology-transfer ventures are usually most effective when led by companies prepared to act as change agents.
- Through its various initiatives, the European Foundation for Quality Management (EFQM) has a key role to play in developing TQM among the members of the European Community, and is a key technology-transfer network.
- The Malcolm Baldrige National Quality Award has had a significant effect on stimulating the adoption of TQM principles in US companies.
- The criteria for the Malcolm Baldrige National Quality Award and the European Quality Award provide useful guidance for an organisation in developing its process of quality improvement.

## Chapter 9 – Creating a Total Quality Improvement Strategy

A strategy for implementing TQI should take into account eight steps:

**Step 1 – Condition Setting**

Before the organisation can formulate and deploy a coherent and relevant TQI strategy, it must assess its needs and attempt to establish a common collective understanding among its senior management team. Condition setting involves three sets of tasks:

- Determining the organisation's current status, both operationally and culturally, i.e., the effectiveness of processes and operational systems on the one hand, and people's attitudes on the other, viz. diagnostic surveys of company strengths and weaknesses, employee attitudes, customer satisfaction and cost of quality.
- Finding out from external sources as much as possible about TQM.
- Focusing top management's perceptions into an awareness of the benefits and mechanisms of TQI.

- Strengths and weaknesses diagnosis should be assessed in relation to external comparative benchmarks – narcissistic organisations soon fade away.
- Senior management's perception of employee attitudes and actual employee attitudes rarely coincide – an attitude survey will show how 'the other half' think.
- Customer dissatisfaction is like an iceberg – mostly out of sight – don't assume your customers are happy and satisfied with your product and service, ask.
- Do not inflate quality costs to shock – their use as a measurement tool depends on clarity, accuracy and credibility.
- The top management team must agree on a common understanding of principles and values as a precursor to setting a strategic direction.
- If the quality council is to be an effective engine of change, it must be led by the chief executive.

**Step 2 – Policy Formulation**

The purpose of formulating a Total Quality policy and strategy is to provide a business direction for ensuring that it is consistent with, and becomes a part of, the organisation's overall business mission. The necessary tasks include:

- Creating a realistic vision and mission for TQI.
- Developing a TQI model for the organisation, integrating all improvement activities into the TQI process.

- Developing a TQI strategy, setting long-term objectives for Total Quality, clarifying short-term priorities, and assigning responsibilities.

- The policy should address vision, mission, goals and values.
- Objectives must have an external as well as internal orientation – customers are more interested in 'what' than 'how'.
- Use objectives as a tool for appraisal and motivation to ensure that quality remains the number one business priority.
- Improvement plans should clarify short-term priorities, resources, competencies, performance measures and comparative benchmarks.
- Performance measures should be set by the managers who are going to make things happen.

**Step 3 – Policy Deployment**

Policy deployment involves the chief executive and the top management team sharing the business direction for Total Quality throughout the organisation, and providing the necessary conditions for implementation, including establishing an improvement infrastructure. This involves such tasks as:

- Cascading the vision, objectives and priorities down through the organisation to initiate a collective sense of purpose.
- Providing the necessary conditions for implementation and cultural change.
- Demonstrating personal commitment to, and involvement in, the TQI process.
- Monitoring the effectiveness of the TQI process and making changes to the strategy where necessary.
- Motivating participants in the TQI process by recognising achievement.
- Continuing to promote all aspects of Total Quality.

- Policy deployment means aligning people to the strategic direction of the business.
- Policy deployment is not a one-off activity – it is a continuing cycle of stimulating action, review, analysis and renewal.
- Policy deployment is a two-way street: goals cascade downwards and the organisation's voice rises upwards.
- Empowerment means providing people with change opportunities, tools and skills to enable them to contribute.

- Education and training starts at the top, i.e., with the chief executive.
- To be effective, training requires a business-related strategy, and an audit mechanism for measuring progress.
- Effective policy deployment is a function of leadership – the key is to develop good habits.

**Step 4 – Improvement Projects**
Initiating the actual improvement process requires the creation of an improvement infrastructure in which the organisation's members can gain experience in problem-solving and participate in change activities. The tasks involved include:

- Creating improvement opportunities throughout the organisation.
- Assigning improvement project responsibilities.
- Forming quality improvement teams.
- Training in problem analysis and solving techniques, and teamwork skills.
- Establishing corrective action and data analysis procedures.

- Initiating quality improvement activity necessitates giving people the means by which they can gain experience in problem-solving, project management and teamwork.
- The first quality improvement teams should be formed at the top of the organisation.
- Quality improvement projects are building blocks to develop a leadership culture to cope with never-ending change.
- Visibility and teamwork are two core factors which shape the nature of a quality improvement project.
- The Quality Council must co-ordinate the project review process and spread the resulting benefits and lessons throughout the organisation.
- Categorising problems into the categories of 'vital few' and 'useful many' can help guide priorities for action.

**Step 5 – Quality System**
The Kaizen strategy of continuous, incremental improvement followed by standardisation of gains, necessitates the development of a company-wide quality system. These provide a systematic framework of reference points from which to map and then internalise the changes brought about by the improvement activities. The develop-

ment of a company-wide quality system in accordance with the requirements of the ISO 9000 series of standards can provide such a framework. Therefore, at the same time as improvement activities are initiated the following tasks should begin:

- Identify the strengths and weaknesses of the quality system against the requirements of the ISO 9000 series.
- Develop appropriate quality assurance procedures.
- Develop a quality system manual.
- Extend the system framework throughout the organisation.
- Establish quality system auditing procedures.

- A quality system meeting the requirements of the ISO 9000 quality system series is a sound basis for managing the standardisation dimension of TQI.
- The purpose of a quality system is to provide a framework of reference points to help cope with the complexity of the organisation.
- The design and development of a quality system should be treated as a project.
- The quality manual should be seen as an atlas mapping out the organisation's process network.
- Formal internal audits are essential to ensure that the quality system remains effective.
- The quality system should be treated as a management information system.

**Step 6 – Process Improvement**

The experience gained in problem-solving should now be extended to improving the capability of actual processes to meet assigned objectives and satisfy customer needs. The tasks involved include:

- Establishing process ownership and responsibility.
- Determining process, product, service and information flows.
- Determining process objectives and associated activity elements.
- Identifying, clarifying, and agreeing with process customers, their requirements, and the performance measures needed to ensure that these requirements are met.
- Identifying with process customers areas for improvement and improvement projects.

- Process improvement is concerned with the capability of individual processes to add value to the overall process stream.

- Process improvement involves three complementary tasks: getting the process under control, process engineering, and process alignment.
- Eliminating causes of process variation concerns service-related processes just as much as manufacturing-related processes.
- Process improvement necessitates analysis of process purpose, ownership, input, output, constraints and impact on other processes.
- Process improvement activities will inevitably affect other process owners sooner rather than later.
- Effective progress begins with elimination of intra-process constraints, and is sustained through a joint commitment by adjoining process owners to eliminate inter-process constraints.
- Inter-process teamwork and joint problem-solving experience is an essential prerequisite to successful process stream improvement activity.

**Step 7 – Process Stream Improvement**
Improvement methodology and teamwork skills can now be applied to improving the capability of whole process streams to add value throughout the entire organisation. This is facilitated through the introduction of:

- Quality Function Deployment (QFD).
- Cross-functional management.

- The goal of process stream activities is to create conditions for effective cross-functional management.
- QFD can be used to align the operational objectives of an entire process stream to meet the requirements of the ultimate customer.
- Trust and experience of the benefits of cross-functional improvement activities are required before QFD can be used effectively.
- External suppliers form part of the process stream, therefore 'co-makership' is an essential component of process stream improvement.
- Process stream improvement takes place in a changing, dynamic business environment – it is not a 'one-off' activity.

**Step 8 – Strategic Benchmarking**
To remain effective, a TQI strategy must be dynamic. This requires a constant renewal of goals, objectives and action plans. Benchmarking

against best-in-class and competitors is an essential part of this renewal process, in order to provide relevant milestones of progress and generate organisational creativity in achieving competitive advantage. The tasks involved include:

- Identifying suitable areas and performance measures for benchmarking.
- Identifying comparable organisations.
- Determining data collection methodology.
- Determining 'performance gap'.
- Communicating benchmark findings within the company, and gaining acceptance.
- Integrating benchmark findings into the continuing policy formulation and deployment process.
- Implementing specific actions and monitoring progress.
- 'Recalibrating' benchmarks.

- Strategic benchmarking is a creative search for ideas to build and sustain competitive advantage in an ever-changing environment.
- Analysis should focus on 'how' as well as 'what'.
- Benchmarking is itself a process of incremental renewal.
- Everyone in the organisation should be encouraged to conduct their own benchmarking activities.
- Organisations must benchmark against the six strategic dynamics of TQM: customer care, competition, leadership, suppliers, people and money.

## Chapter 10 – Becoming the Best

- There are six different levels of involvement in TQI adoption:
  - Uncommitted
  - Drifters
  - Tool pushers
  - Improvers
  - Award winners
  - World-class

- These levels represent the characteristics and behaviour which organisations display in relation to TQI.

# Chapter 1
# What is Quality

## Introduction

'Quality' is one of those words of which everyone has an intuitive understanding but most people are hard pressed to define; "I know it when I see it." As Price[1] comments, the word 'quality' must have been used more in the last ten years than in the preceding ten centuries, yet the more we hear it the more confusing its meaning seems to become. So what is 'quality'? The basic answer is simple, perhaps deceptively so.

Quality is what gives *complete customer satisfaction.* This is true for every customer whether it be industrial buyer, local authority, government department, or for ourselves with regard to the products and services which we purchase every day of our lives. Some customers' needs are sophisticated, others much less so – but all will judge the quality of what they buy by how well their needs and expectations are satisfied both in the short and long term. Many organisations are now talking in terms of delighting customers by giving them more than they expect and through this keeping and winning customers.

Superior-performing companies are using quality to win market-share, enhance their competitive position and improve profitability. Having made quality the central driving force of their operations, an increasing number are riding the momentum generated to become world-class performers. These companies have recognised that the key concept of quality is to meet customer requirements, on time, every time, 100% of the time.

One hundred per cent of the time? Surely, that is at best optimistic, and at worst just another 'motherhood' statement – the kind of overblown rhetoric that advocates of TQM seem to be so keen on. Consider the following two comments by James Houghton, chairman of Corning Glass:

> *We had a Corning salesman who went out a couple of years ago very proudly and said to a customer: "We are now operating at 1%*

> *AQL (Acceptable Quality Level), isn't that wonderful? That means 99% good." The customer turned to him and said, "Fine, but now we're going to call that 10,000 parts per million, and we're going to work it down from there." A very different way of looking at it.*
>
> *Frankly, 99% is not good enough. You know what 99% is? Ninety-nine per cent is 20,000 pieces of lost mail per hour. Ninety-nine per cent is two short or long landings per day at Heathrow. Ninety-nine per cent is 5,000 incorrect surgical procedures per week. Ninety-nine per cent is 200,000 incorrect prescriptions per year.*[2]

Customer satisfaction: think about it.

## Defining quality

Satisfying the customer, delighting the customer, and winning customers, are all typical of the phrases which encapsulate organisational definitions of quality. For example, the Ford Motor Company of Europe:

> *Quality is defined by the customer. The customer wants products and services that throughout their life meets his or her needs and expectations at a cost that represents value.*[3]

The internationally accepted formal definition of quality (ISO 8402) is:

> *The totality of features and characteristics of a product or service that bear on its ability to satisfy stated or implied needs.*[4]

Two other commonly quoted definitions of quality are worth mentioning; those of two of the leading authorities on the subject of TQM – Crosby and Juran.

Crosby[5] defines quality as conformance to requirements, meaning the degree of conformance of all a product's features and characteristics to the product requirements specification and the stated requirements of users. This definition is often quite useful for facilitating the internal customer/supplier relationship, particularly when they involve non-manufacturing functions in the process of quality improvement. Internal customers should, in conjunction with their suppliers, define and agree a clear set of requirements.

Juran[6] uses the term 'fitness for purpose' because he maintains that the ultimate user's evaluation of quality is based on whether the product is fit for use, and not on whether it conforms to the specification requirements (of which the user may not be aware in the first place). The word 'quality' has traditionally been used as an imprecise adjective to imply a high degree of excellence, grade, level, technical sophistication, or some other distinguishing attribute, such as 'goodness, or luxury, or shininess or weight' (to quote Crosby[5]). For example, just look through any argument or case for funding, in relation to keeping open some school, unit, institution, department, etc. – the discussion of some action or other; the promotion of an individual; the advertising of some product and service – and see how many times the word 'quality' is used and in what context. It is clear that much of the usage of the word quality is misleading and serves to confuse and obscure its true meaning with the myth that: quality is a luxury; quality is expensive; quality is something for which you (the customer) are prepared to pay a premium, because 'high' quality incurs an extra cost to the manufacturer or service provider. In short, quality is often confused with 'grade'.

The traditional perception that quality (when the word 'grade' should have been used) is costly and that you only get what you pay for has led to the notion of a trade-off between cost and quality. This is the root cause of many of the common misconceptions regarding quality. The optimum quality level concept, which is portrayed in many production/operations and strategic management texts, serves to reinforce this view. Fortunately many people are now recognising the flaws in this notion. There is little doubt that customers will pay a premium for additional product and/or service features or characteristics and design technology to satisfy their needs, but this is grade, not quality. A quality product and/or service is one which conforms to stated requirements and satisfies the customer in terms of its use and value for money.

This notion is not limited to 'industry'; academia can also be guilty. Parasuraman et al[7] reiterate the cry of many academic writers and researchers that quality is an elusive and indistinct construct. They suggest that few academic researchers have attempted to identify and model quality because of the difficulties involved in isolating and measuring the construct. Those that do, often bypass definitions and use unidimensional measures or theoretical models to explain the concept.

There has been considerable speculation in the academic literature on the relationship between price and quality. Researchers such as McConnell[8] and Jacoby et al[9] postulate a unidimensional relationship as shown in Figure 1.1, i.e., high price signifies high quality and vice versa. Such research findings appear to provide empirical evidence which reinforce the notion of a trade-off between cost and quality. On the other hand, some researchers are convinced of the existence of such a trade-off but appear to be frustrated at their inability to find theories capable of supporting it, e.g., Geistfield.[10] After conducting a survey of 145 different products, Gerstner[11] expressed surprise as he expected to find in a 'perfectly functioning' market a strong, positive relationship between product quality and price, but the results would not bear out his hypothesis.

Fortunately, the evidence of the Profit Impact of Market Strategy (PIMS) database as reported by Buzzell and Gale,[12] Phillips et al[13] and Luchs[14] among others, appears to have certainly helped, in academic circles, in laying to rest the ghost of price/quality trade-off. These writers find no direct correlation between relative quality and higher direct cost, and conclude that the opposite is often the case. Subsequent studies have established that quality can be equated to customer perceived value, and that companies with large market-shares are those with high relative customer perceived value, i.e., quality, whereas companies with small market-shares are those with low relative customer perceived value, e.g., van Nievelt.[15]

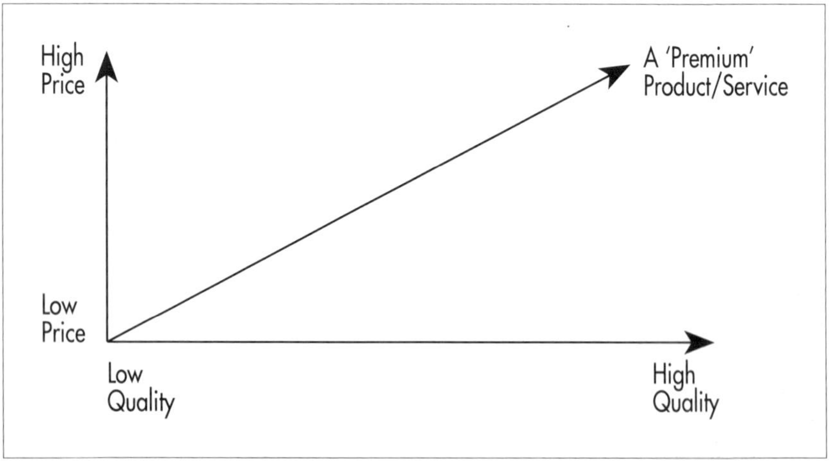


**Figure 1.1** Unidimensional price/quality relationship

Some researchers have developed theoretical models of alternative strategies which are applicable to differing market conditions and product life-cycle stages so as to optimise profitability. Walsh,[16] for example, has constructed such a model relating specifically to alternative quality management strategies. The underlying assumption appears to be that certain product attributes, such as quality, are either optional or capable of being fine-tuned in accordance with changing market conditions, e.g., when faced with an over-mature market or a product in the decline phase of its life cycle, reduce the 'quality content' as part of a 'quit' strategy to optimise profit contribution. This assumption overlooks two facts: First of all, a considerable part of Europe's manufacturing base is in industries which have reached (in theory) the decline stage of their life-cycles. Secondly, demanding customers will no longer allow suppliers to determine product and service quality arbitrarily (it either meets their needs and expectations or it does not – there is no sliding scale of fitness for purpose). The following example from our research provides further evidence which questions the validity of such models:

A company launched a new business venture which involved the manufacture of tin cans, a generic product which could be said to be well into the decline stage of its life-cycle (the very point at which the accepted wisdom of strategic planning states that participants should withdraw from the market, or at least optimise the remaining profit potential by reducing optional product attributes such as quality). Furthermore, the company entered a market plagued by over capacity and demanding customers. But by collaborating with its customers through joint planning activities to produce what could be theoretically termed a 'high quality' product at a competitive (not premium) price, the venture went on to earn, with boring regularity, at least 40% return on net assets employed.

Quality is most certainly not an optional product and service attribute. All the evidence is that it is now the main business strategy of an increasing number of enlightened organisations, some of which have acquired world-class status. These organisations have not only survived, but have actually increased their market-share by making quality their number one priority.

Quality can only be achieved if everyone within the organisation believes in it – the constancy of purpose referred to in the first of Deming's[17] fourteen points. To knowingly produce and deliver non-conforming goods or services is to live a lie. No organisation can live a lie and survive in the long-term – the truth will inevitably emerge. Handy,[18] a provocative writer on management issues, suggests that quality is the organisational equivalent of truth: *'Quality, like truth, will count in the end'.*

This chapter now examines some of the relevant issues of quality management, describes in brief the evolution of various approaches to the control and assurance of quality to the ultimate goal of TQM, and explains the concept of Total Quality Improvement.

## Quality management

The pursuit of quality is not easy. It requires:

- Motivation for change (both of the organisation and individuals).
- Commitment.
- Direction and leadership.
- The right people.
- Education and training.
- The right systems, tools and techniques.

All of these are management issues. Quality management is that aspect of the overall management function that organises and directs all quality-related activities.

The *quality* of a product and/or service is both a goal and a tangible outcome, whereas, *quality management* is a dynamic set of activities to achieve the goal and the outcome. Quality management is about doing the right thing, right first time and every time to achieve complete customer satisfaction as efficiently and as cost-effectively as possible.

## Parameters of quality management

There are three parameters which impact on the set of activities called quality management: these are customer perception, competition, and business performance (Figure 1.2).

**Customer perception**

There is more to customer satisfaction than mere 'conformance to requirements' or 'fitness for use'. Whilst the supplier's ability to

**Figure 1.2** Three parameters of quality

deliver goods and services on schedule to meet customer-agreed specifications is crucial, even more important is how the customer perceives the quality of the product or service, and the supplying organisation itself, including the behaviour of its staff. Customer-perceived quality is influenced by the customer's image of the business and people as well as his image of the supplier's products and services. A distinction is sometimes made between quality in fact and quality in perception. Townsend,[19] for example, defines quality in fact as 'doing things as they should be done' and quality in perception as 'doing what needs to be done'. But the distinction is somewhat artificial – in the provision of services, for example, the two are inseparable.

From his own perceptions, and from promises made by the supplier, the customer develops expectations. When the order is fulfilled and/or service delivered and experienced, the customer gets some experience of whether or not the supplier lives up to these expectations. Image, expectations and experience directly influence customer-perceived quality and therefore customer satisfaction, all of which are elements of a company's reputation for quality. This is achieved through creating a mix of customer benefits based on a combination of design quality, product quality, sales quality and service quality (Figure 1.3).

Teboul[20] provides a useful visual analogy (Figure 1.4): customer segments (more or less homogeneous clusters of customers) targeted for satisfaction are represented as a circle; the product or service offering as a square. The task of marketing and design is to position the square on the circle to achieve the best possible fit in terms of meeting product and service quality-related needs and expectations – 'squaring the circle'.

Several researchers, including Gummesson[21] and Zeithamal et al[22] maintain that there is always a perception distance between supplier and customer. The gap has several aspects, including:

- The gap between the buyer's perception of his needs and the supplier's perception of his needs.
- The gap between the supplier's perception of the offer and the customer's perception of the offer.
- The gap between the customer's experience of what he received and how he received it and what the supplier thought he delivered.

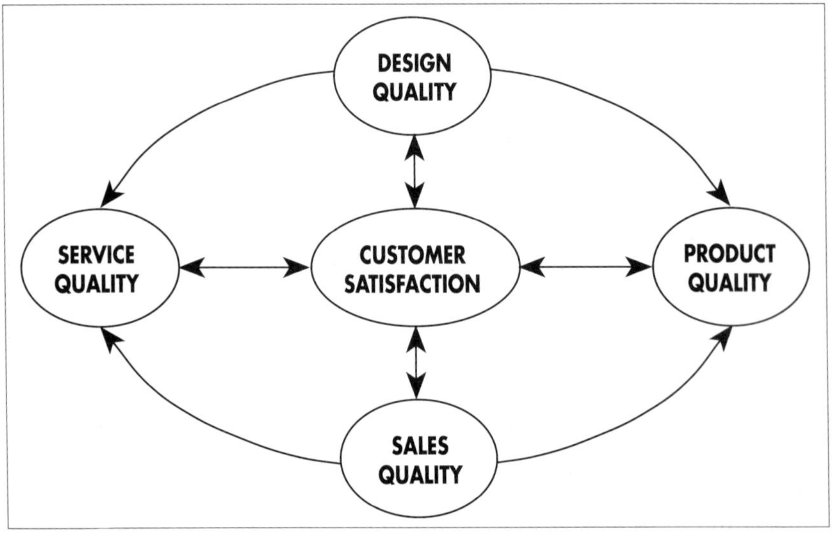


**Figure 1.3** Customer satisfaction concept

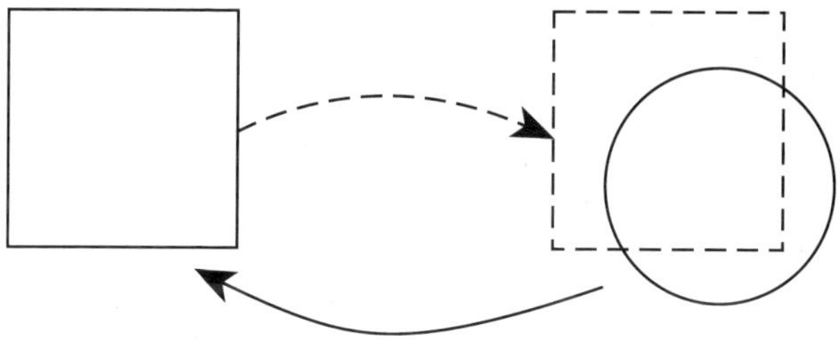


**Figure 1.4** Squaring the circle of customer satisfaction (Source: J. Teboul, *Quality Dynamics*)

Quality management is concerned with bridging, or at the very least narrowing the gap.

**Competition**

Competition is fierce, and getting fiercer in today's turbulent business environment. Every market sector is attracting new competitors, product life-cycles are shortening, 'unique selling factors' of products and services are becoming indistinguishable – market segments fragment, new specialisms abound, product and process innovation intensifies, and market-entry costs soar.

Everywhere there appears to be pressure and chaos, and within the chaos, business organisations attempt to gain competitive advantage. Competitive advantage exists when customers value an organisation's products or services in preference to competitive alternatives.

In the 1980s, achieving sustainable competitive advantage became the cornerstone of marketing strategy. The two most commonly proposed approaches were product differentiation (usually by building in one or more superior 'quality' attributes to achieve a high market price), and cost leadership (by maximising the efficiencies of volume production and of the learning curve). Conventional wisdom postulated that these two were mutually exclusive strategies, but this hypothesis was based on the notion of the cost versus quality trade-off discussed earlier. By the middle of the 1980s, even the most obtuse of marketing strategists began to realise the flaw in this argument. The evidence was before their eyes in terms of significant losses in market-share and profitability to their Japanese competitors who combined product differentiation and cost leadership through their application of quality management. In effect, the Japanese had 'moved the goal posts'. This strategy is now being followed not only by the Japanese and their Pacific Rim neighbours, but by an increasing number of American and Western European organisations who are exhibiting superior performance in the market-place.

Feigenbaum[23] says that today's effective international competition is a combination of competition in its traditional and visible form (product versus product and service versus service) and an equally powerful, but less visible form of competition involving companies' skills in quality management.

Deming[24] and Juran,[25] now internationally celebrated TQM gurus, were advocating the importance of quality management as a competitive weapon over 40 years ago. But in what was basically the sellers' market of the post-war world, their message went largely unheeded – except by the Japanese. Although the importance of product and/or service quality as a competitive weapon is paramount, orders cannot be won on quality alone. The major effect of the shift in the market paradigm is that acceptable or good quality has become a minimum requirement for market entry. Hill[26] argues, quality has now become a 'qualifying' criterion in the market-place.

Kotler et al[27] observe that a noticeable feature of the Japanese market entry strategy is the way they link price, quality and service as three distinct attributes of their offering compared with the competition. They cite the example of how Japanese machine-tool manufacturers penetrated the American market. The Japanese consistently priced their products below the prices charged by American manufacturers. Initially, the lower price was intended to attract the attention of American customers. However, the Japanese manufacturers also recognised that price alone would not sell their machine tools – thus they placed tremendous emphasis on product quality, reliability and precision. Recognising that down-time had to be minimised, they also offered an immediate repair and maintenance service as part of the product package. In effect, they developed a total package of value-adding benefits at no extra cost to the customer.

Competitive advantage now depends on giving customers more than they expect and enhancing their perceptions of the value content of the product or service relative to competitive offerings, without charging a premium for this. The superior company will always focus on optimising efficiency in quality, cost and delivery to improve competitiveness.

Quality management is concerned with achieving and sustaining competitive advantage.

**Business performance**
Each business organisation, no matter whether it provides manufactured goods or a service, is engaged in the transformation of inputs by means of a network of processes into saleable outputs. Every transformation process involves harnessing the inputs of people, material, equipment, energy and money as efficiently and as cost-effectively as possible (Figure 1.5). The input resources are also scarce and expensive and it is crucial that they be employed without undue wastage.

In general, all activities within an organisation fall into one of two categories: those that add *value* and those that add *cost*. The fact that profit is generated solely by adding value to inputs is often overlooked. In order to add value to the transformation process(es), and hence to the product or service, it is necessary to reduce and ultimately eliminate waste in all its various forms:

- Waiting.
- Motion.
- Transportation.
- Inventory.
- Defects.
- Excess production.
- The process itself.

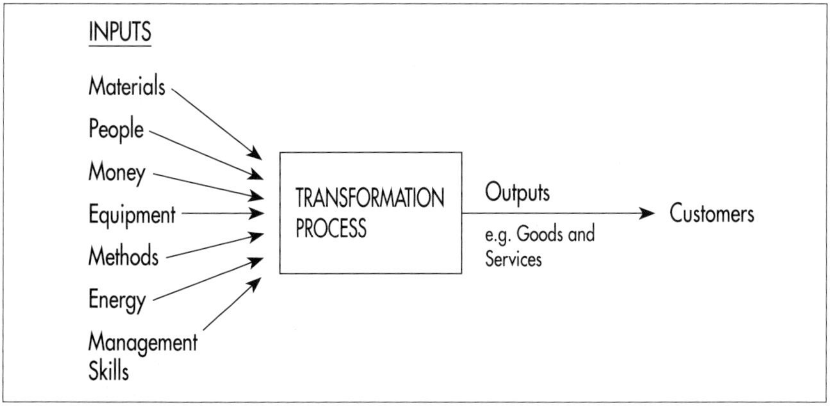


**Figure 1.5** The transformation process

The waste elimination cycle is:

- Identify waste.
- Find the cause of waste.
- Make improvements.
- Standardise the improvements made in order to hold the gains and to transfer the improvement to other processes.

More than one study has calculated that quality costs can account for anywhere between 5% and 25% of the annual sales turnover of a business organisation – much of this cost is avoidable. These losses, caused by the continuation of non-added value transformation processes, are equally true of both service and manufacturing organisations and represent a staggering figure. Just consider what the effect on the combined Gross National Product of Europe's community of nations would be if most of these losses were eliminated.

Quality management is often perceived in company boardrooms as collection of manufacturing-related techniques which are best delegated to the quality assurance department. In such organisations, a distinction is made between 'on-line' and 'off-line' activities. On-line activities are seen as those production activities which convert raw material or assemble components into the finished product, while everything else is seen as off-line and, therefore, intangible. As a result, quality management activities are focused solely on the 'on-line' activities. In reality, any organisation is a network of interrelated processes, both technical and administrative.

There is no such thing as an off-line quality management activity – all processes have an impact on the organisation's capability to add value.

Quality management is concerned with an organisation's capability to add value through its transformation processes.

## Quality management is company-wide

Further consideration of the definition of quality management suggests that the activities of quality management are no different from the activities of management *per se*. ISO 8402[4] states: "Quality management includes strategic planning, allocation of resources and other systematic activities for quality such as quality planning, operations and evaluations."

Remove the word 'quality' from the above definition of quality management and it is difficult to perceive how the concept of quality management could be considered as separate from the general concept of management itself. Indeed, Crosby[5] goes so far as to claim that there is no such thing as a quality problem, there are only management problems.

Ishikawa[28] and Feigenbaum[29] have helped to widen the concept of quality as a management discipline, although both use the term 'quality control' in preference to 'quality management'. Feigenbaum expands the concept of quality control to cover all the managerial control functions of the organisation and suggests that quality is, in essence, a way of managing an organisation, what he calls 'Total Quality Control' (TQC). Ishikawa prefers the term 'Company-Wide Quality Control' (CWQC), saying that this is a better description of the Japanese approach to the management of quality.

These two terms reflect a subtle difference in emphasis. Feigenbaum intends TQC as a framework, encapsulating the four principles of control, i.e., setting standards, appraising conformance to these standards, acting to ensure standards are met, and planning for improvement in standards, to provide an integrated operating system for managing all functions within the organisation. Ishikawa[28] uses the term CWQC to describe the involvement of all functions within the organisation in managing quality, and suggests the phrase 'the next process is the customer', the emphasis being on the interactions between all functions within the organisation in order to optimise the value which is added to the product or service being offered to the ultimate user.

No matter what abbreviation is used, whether it be TQC, CWQC or TQM, Western business organisations are increasingly developing their knowledge of quality management and its application.

## The evolution of Total Quality Management

During the past two decades simple inspection activities have been replaced or supplemented by quality control, quality assurance has been developed and refined, and now the most progressive companies are working towards the concept of TQM.

It should be noted that the terms are used here to indicate phases in the evolution of the management of quality (Figure 1.6 is adapted from Dale and Plunkett[30]) and do not necessarily correspond to the usual definitions, e.g., ISO 8402.[4]

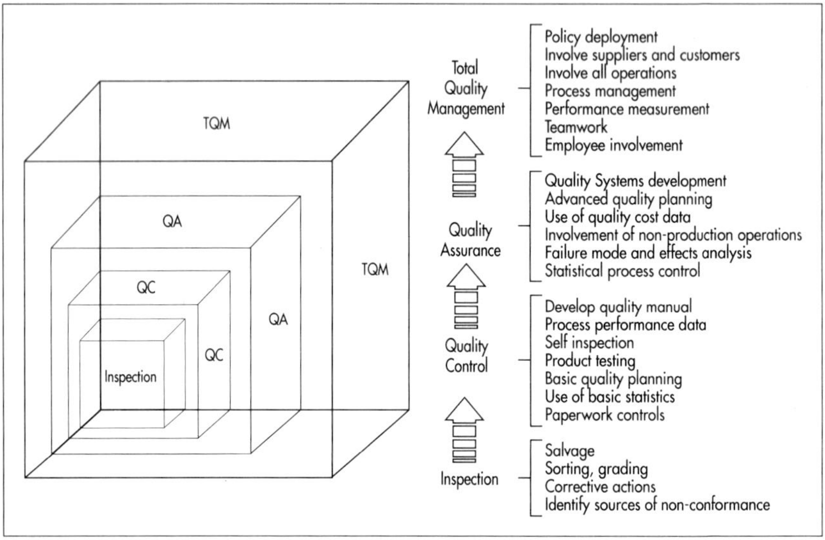


**Figure 1.6** Evolution of TQM

## Inspection

Under a simple inspection-based system, one or more of the characteristics of a product or service offering are examined, measured, or tested, and compared with the specified requirements to assess conformity. The system is applied to in-coming goods, manufactured components and assemblies at appropriate points in the manufacturing process and at the finished-product stage, before despatch to the customer or before/during delivery of the service. It is usually carried out by staff who are employed specifically for the purpose. Goods or products which do not conform to specification will be separated out and may be, for example, scrapped, reworked, blended off, reground for subsequent reprocessing, passed on concession, or downgraded. An analysis of a typical distribution of the key dimensions of a component from a batch of product will indicate to the user the degree to which this type of sorting has been carried out (Figure 1.7).

These types of activity are obviously more difficult to undertake in service situations, and in any case may be impractical as the customer is usually directly involved in the transformation process. In the delivery of services, it is often the case that the first time an organisation is aware that something has gone wrong is when the customer complains. The nearest that services get to downgraded

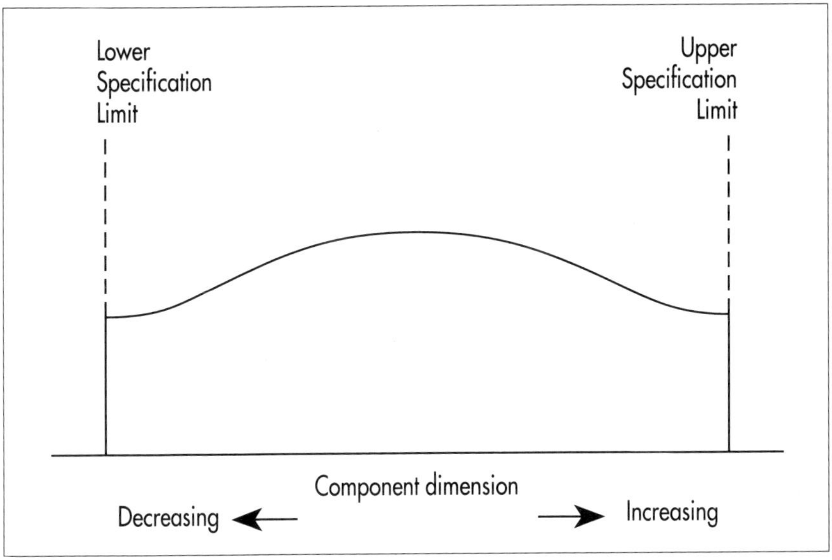


**Figure 1.7** Sorting components

products is in the situation when there is no suitable alternative, and due to continually non-conforming services the customer lowers his expectations (regular air and rail travellers can speak from personal experience).

The system employed in inspection is an after-the-fact screening process to catch defects before they are shipped to the user, and has no prevention content other than perhaps detecting faults in procedures and systems, identification of suppliers, operations and/or workers producing the non-conforming products/services. This is a wasteful usage of resources because the product or service has had to be produced and delivered before it can be checked. Simple inspection-based systems are usually wholly in-house and do not directly involve suppliers or customers.

## Quality control

Under a system of quality control one might expect, for example, to find in place a design control system, raw material and intermediate stage product testing, some self-inspection by operators, logging of elementary process performance data, and feedback of process information to operatives and first-line supervisors. Whilst the main mechanism for preventing products which are outside specification from being shipped to customers is again screening inspection, improved quality control measures lead to better process control and fewer incidences of non-conformance.

Inspection and quality control are based on detection type activities and, in the main, are focused on the finished product rather than the process and, in the case of service organisations, on the front-end delivery of the service, e.g., customer-care programmes, rather than the up-stream processes, methods and systems (Figure 1.8 – from Ford Motor Company[3]). In the main, the stages of inspection and quality control are short on scientific analysis, planning and measurement. They also tend to be focused on compliance to specification.

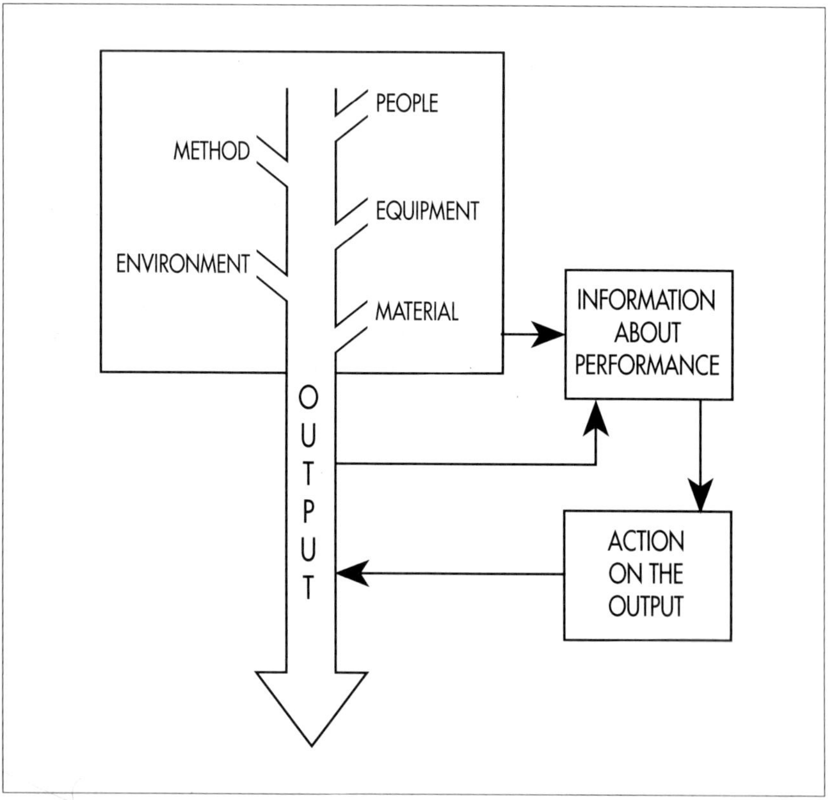


**Figure 1.8** A detection-based quality system

**Quality assurance**

Quality assurance involves all the planned and systematic actions which are required to give customers adequate confidence that a product or service will satisfy given stated requirements throughout its life cycle. Examples of additional features acquired when progressing from quality control to the quality assurance phase of quality management are, for example, a more comprehensive and pro-active quality system, gathering and use of quality costs data, use of Statistical Process Control (SPC), Failure Mode and Effects Analysis (FMEA), mistake proofing systems, and advanced quality planning. Above all, one would expect to see a shift in emphasis from mere detection towards prevention of non-conformance. A typical prevention system is shown in Figure 1.9 (Ford Motor Company[3]). It can be seen that action is focused on the process itself in terms of preventing non-conforming products at source, and also on the design of the process and the product or service to make it more robust in the face of process variation.

**Total Quality Management**

TQM means the application of quality management principles to all aspects of the business on a company-wide basis. This requires a strategic commitment by the senior management team and means fundamental changes in behaviour, attitudes, values, culture, organisational structure and work practises. As an organisation develops its approach towards TQM, there is an increasing emphasis on continuous improvement and human relations. An organisation can drift through the stages of inspection, quality control and quality assurance by:

- Responding to customer pressure to obtain quality system registration, e.g., ISO 9000 series.[31]
- Introducing, often as a contractual requirement of the customer, certain quality management tools and techniques.
- Using the experience brought from other organisations by new managers and technical specialists.

This is not the case with TQM, it needs a breakthrough and this can only be initiated by senior management in the form of a strategic decision.[6]

With TQM the principles of quality management are applied in every branch and at every level of the organisation. Individual

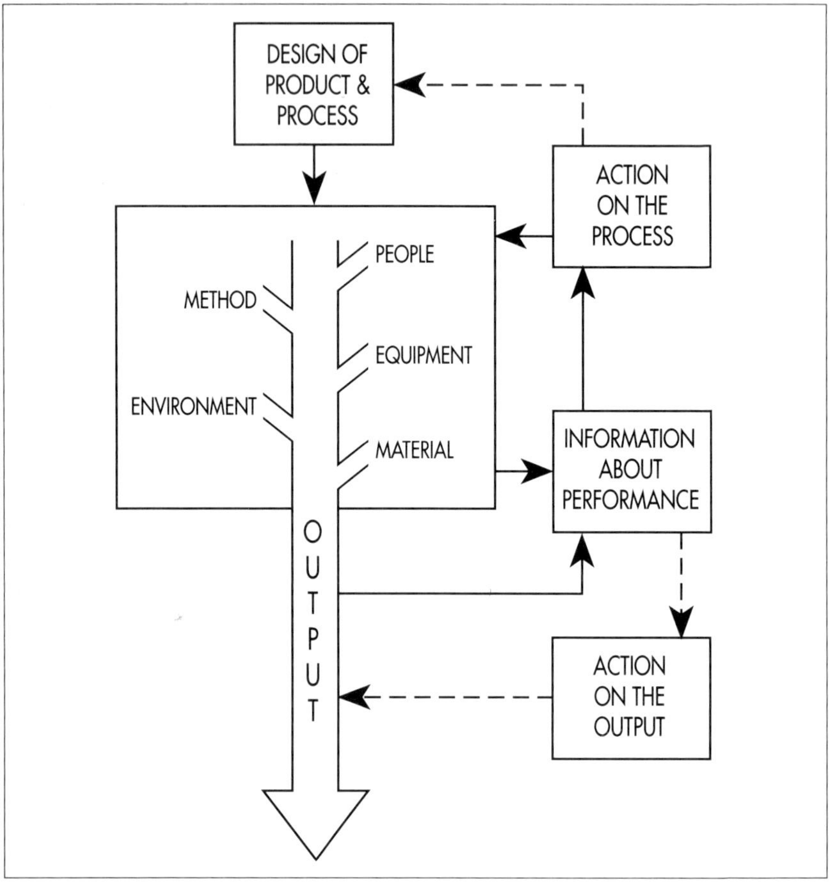


**Figure 1.9** A prevention-based quality system

departmental systems, requirements and standards may be no higher than for the quality assurance phase of quality management, but they will pervade the whole organisation including all functions and business activities. The spread of the TQM philosophy would also be expected to be accompanied by greater sophistication in the type of quality management tools and techniques which are applied.

The key features of TQM include:

- Each person at every level of the organisation being prepared to accept *total ownership* and responsibility for quality improvement within their own sphere of influence.

- A variety of methods will be employed to involve, motivate, and imbue people at all organisational levels with the philosophy that *improvement is a way of life.*
- Quality improvement activities will extend beyond the boundaries of the organisation to include *partnerships with suppliers and customers.*
- Widespread application of the concept that *everyone in the organisation has both customers and suppliers.* Customers will define precisely to their suppliers what their requirements are and these suppliers will endeavour, as a personal commitment, to meet their customers' needs and expectations. Key internal performance measures are used to ensure that suppliers focus on meeting customer requirements. If all internal customers are being satisfied, then progress is being made towards the satisfaction of external customers.
- A series of measures will be employed by the organisation to monitor improvement, assess the level of customer satisfaction and benchmark *key performance indicators.*
- *Employee involvement and participation* in the business, and their development, will be pursued and every person's skills and abilities will be fully utilised.
- A variety of *teamwork* activities will be employed to facilitate continuous improvement. This involves teams working with teams to resolve deep-rooted problems.
- *No barriers* between departments and functions.
- *Simplication and standardisation* of work practices.

## Kaizen means continuous improvement

Imai,[32] amongst others, emphasises the concept of *Kaizen* as the key to Japan's competitive success. The word Kaizen simply means improvement – not 'quality improvement' as it is often translated. We first came across the Kaizen concept in 1985 on one of our first visits to the then new Nissan Motor Manufacturing (UK) plant in Washington, Tyne and Wear. During the course of that visit we received several friendly, but firm, reprimands from our hosts for our initial insistence on calling Kaizen 'quality improvement'. "No, no!" they said, "Kaizen means more than just quality improvement! It is concerned with every aspect of our operation. We do not single out quality for special treatment. Our aim is to improve everything NMMUK does." It was then that we realised the full impact of the

Kaizen concept, that it is *total* and not restricted to product and service quality alone.

Kaizen is the epitome of continuous improvement. Kaizen means small changes in the status quo as a result of on-going efforts. Kaizen is a gradual, incremental process. By listening to the voices of the internal processes, the customer, and the interfaces between customer and supplier, the Kaizen strategy is directed at optimising available resources to increase added-value and customer satisfaction. This is a radically different managerial approach from that traditionally taken in the West: periodic attempts to make big, dramatic breakthroughs, which sometimes succeed but more often do not. Imai[32] labels this breakthrough approach 'innovation'.

Innovation strategies are usually concerned with the implementation of new technology or systems, e.g., Computer-integrated Manufacturing, Computer-aided Design/Computer-aided Manufacturing, Manufacturing Resources Planning, Flexible Manufacturing Systems, and new product development which promise an enhanced return on investment. But the 'one-shot' innovation mentality can apply equally as well to the kind of high-profile corporate improvement programmes typified by slogans such as 'zero defects', of which Crosby's fourteen step programme[5] is a prime example. These are usually initiated with a high-level declaration of commitment to the new vision, e.g., zero defects, and pursued with great energy and vigour, resulting in a series of short-term gains, until the programme reaches a plateau, followed by decline and erosion of all the gains. The cycle looks something like the one outlined in Figure 1.10 (from ICI Chemicals and Polymers).

What a waste of energy and resources! What a 'roller-coaster' of emotion and what a damper on morale! What a pity!

Imai[32] depicts the pattern a little more formally: the innovation strategy is supposed to bring about a staircase progression as shown in Figure 1.11. The actual progress is, however, more like that illustrated in Figure 1.12, because the gains of the innovation are subject to a steady deterioration unless continuing efforts are made to maintain and then improve on them as indicated in Figure 1.13.

We have observed that organisations making small changes are more likely to hold the gains than in the case where large-step changes are made where almost inevitably, the system is likely to revert to the status quo.

Whenever an innovation is made, standardisation is required to hold the new performance level. Central to the Kaizen strategy is a

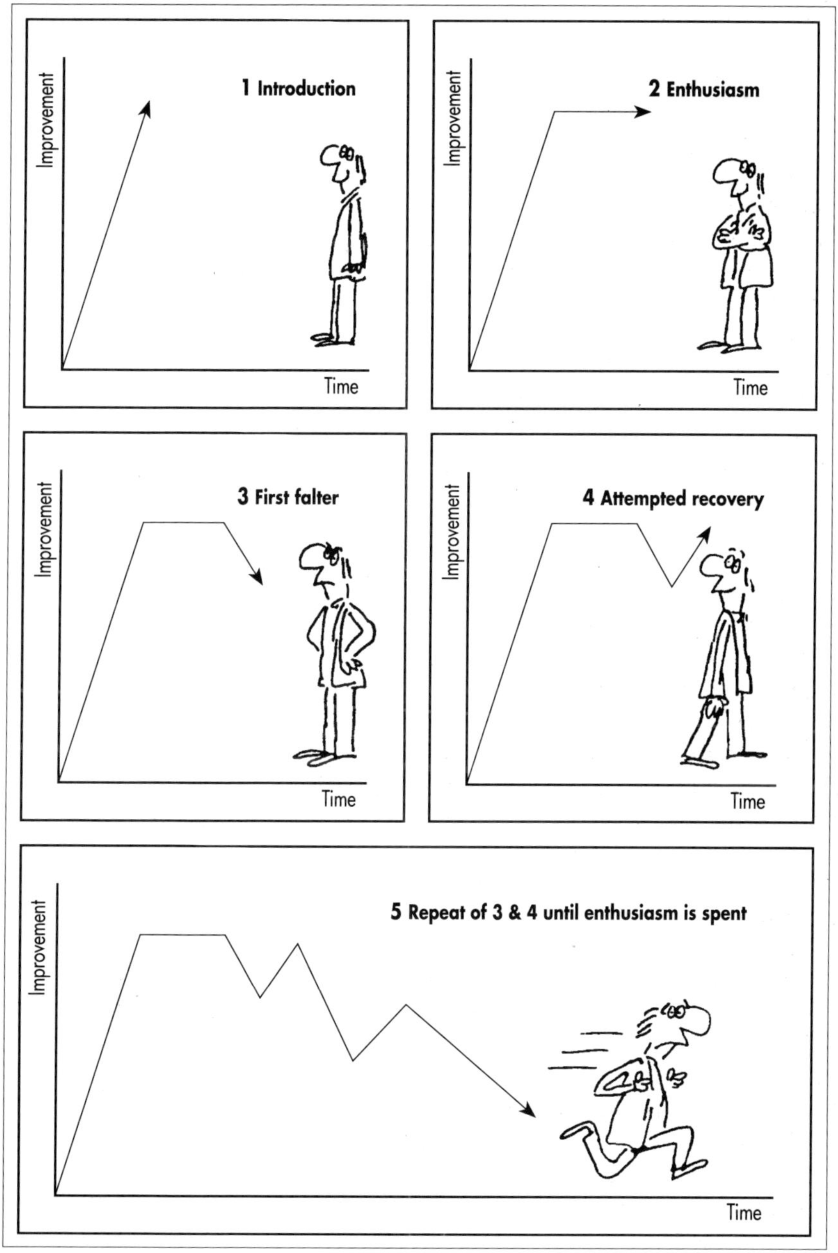


**Figure 1.10** A 'one-shot' improvement programme (Source: ICI Chemicals and Polymers – cartoon by Matt)

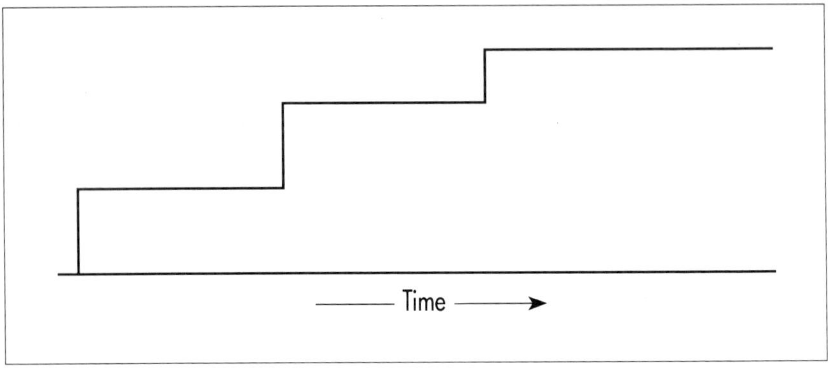


**Figure 1.11** Ideal pattern for innovation

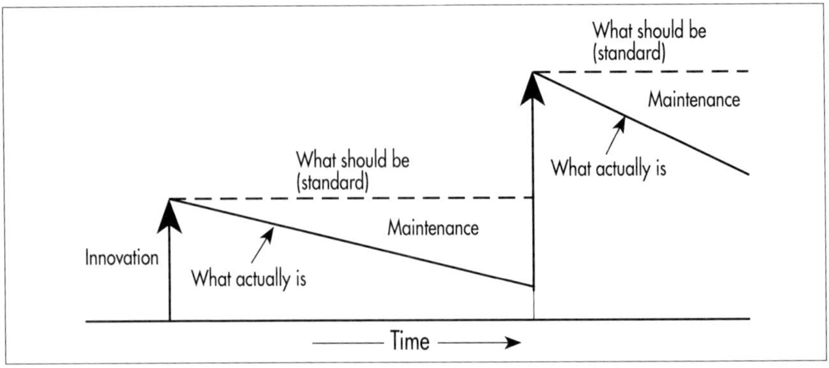


**Figure 1.12** Innovation alone

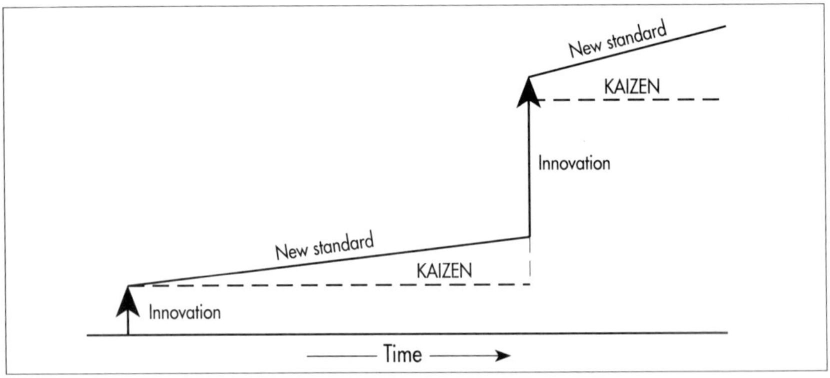


**Figure 1.13** Innovation plus Kaizen (Source: M. Iami, *Kaizen: The Key to Japanese Competitive Success*)

constant effort not only to maintain but also improve standards and the process of standardisation (Figure 1.13). Kaizen strategists see standards as stepping-stones with one standard leading to another as improvement efforts continue. This incremental process is a fundamental basis for managing quality improvement. The Japanese companies view standardisation as a key element in their improvement philosophy (see Chapter 2).

## Total quality improvement

TQM is an integrative framework for managing the business organisation to achieve the goal of complete customer satisfaction. It is also a philosophy which requires radical changes in managerial thought and action. This is a journey and not a destination, and has no ideal state – the current situation in any process can always be improved upon. A process of continuous improvement is the enabling mechanism for TQM.

This never-ending process of improvement requires the creation of an organisational culture which recognises the existence of problems, and which enables everyone within the organisation to diagnose and solve problems. As problems usually spread beyond departmental boundaries, a cross-functional approach to managing improvement is essential. But this is not all. To succeed, a real understanding of the customer's wants and needs, clear objectives, leadership and education are also required. We have called this Total Quality Improvement (TQI).

Why Total Quality Improvement? Well, let us take each word in turn starting with *Improvement*, for which by now we hope the need is clear – corporate survival depends on never-ending improvement. But why *Quality*? Surely improvement as epitomised by the Kaizen concept is not limited to quality – after all, there is no such thing as a quality problem, so the word quality must be redundant. Not so. We believe that the concept of quality as organisational trust, truth, and concern and care for people is the outstanding emotive force which can unite everyone within the organisation. Neither should it be forgotten that quality, in the proper framework, can be used as a common corporate language throughout an organisation, and this helps the unification of effort and common direction. It is *Total* because it is an enabling mechanism which is not only company-wide, but pervades the entire environmental setting and binds the organisation with its customers and its suppliers.

We visualise TQI as a three-dimensional shape as illustrated in Figure 1.14. Along two dimensions are the Kaizen forces of improvement and standardisation, along the third are what we term the strategic dynamics of TQM (customer care, competitors, leadership, suppliers, people and money) affecting progress towards the ultimate goal of a TQM culture and complete customer satisfaction. The dimensions of improvement and standardisation are mainly internal in focus, whilst the strategic dynamics of TQM tend to be external factors which may provide the stimulus to advance the process and act as milestones of progress. The point is that improvements will have an internal impact, but unless judgements are taken in relation to some external orientation, an organisation will not know their effect in terms of improved customer satisfaction or competitive advantage.

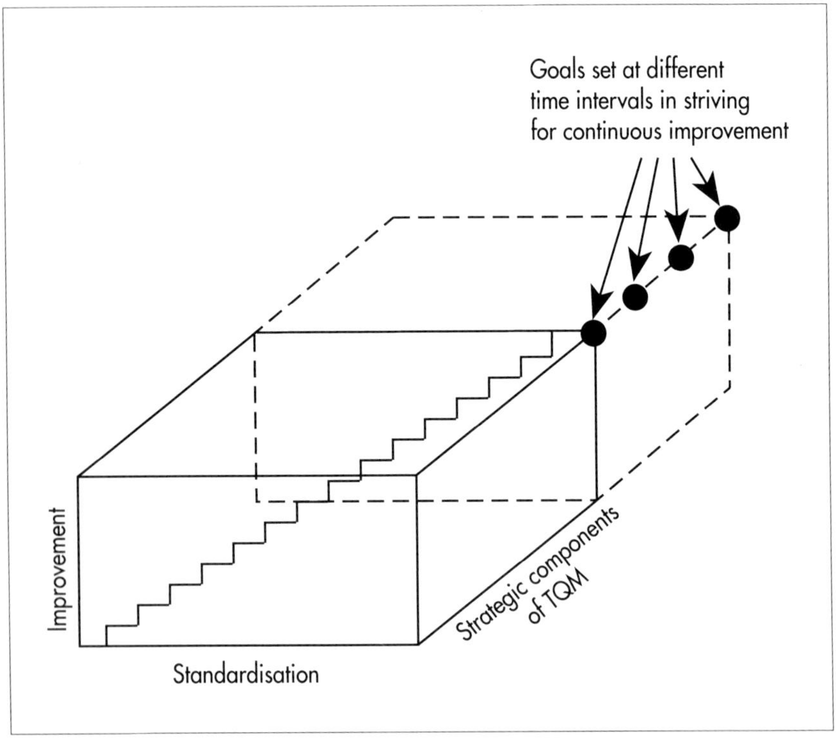


**Figure 1.14** Total Quality Improvement

**Improvement**

Quality improvement is a continuous process. In essence, it is concerned with enhancing the capability of an organisation's network of processes to add value to products and services in order to increase customer satisfaction. We make no apology for labouring this point: it is fundamental to understanding the concept of TQI.

As Conti[33] points out, a stream of processes is not just the sum of the processes. Interactions among the processes within the streams, and between these and the surrounding environment, means that the stream behaves as a global, complex, dynamic system, not just the aggregate of separate operational activities. As the stream of processes lengthens and becomes more complex, the emphasis for reducing non-added value activity, i.e., waste, shifts from technical and methodological factors to management-related factors (Figure 1.15). For example, the weak points are at the boundaries of functional departments where the various interfaces of processes and activities occur, and these need to be identified and carefully managed. Cross-functional management activity, with the management

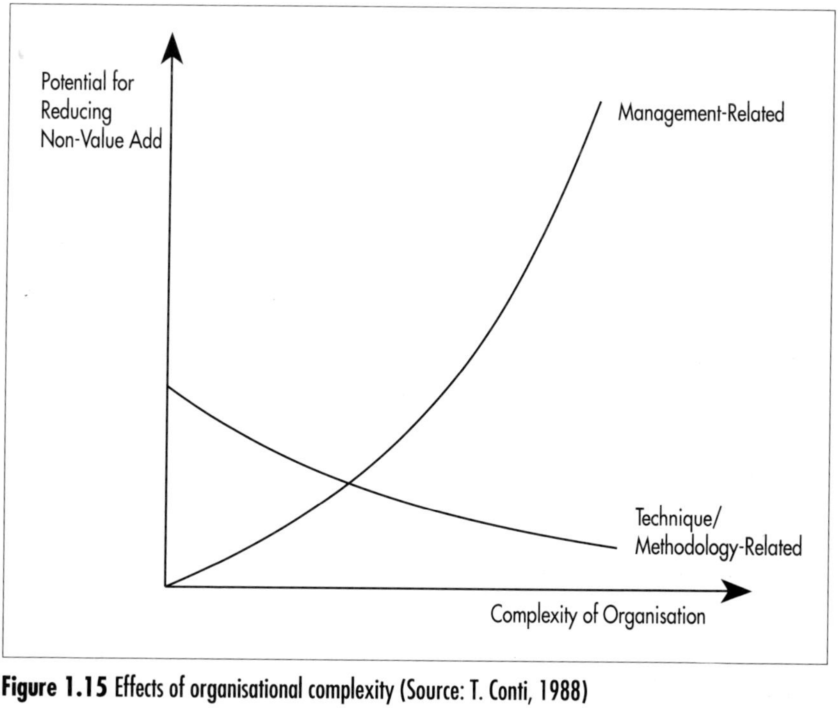


**Figure 1.15** Effects of organisational complexity (Source: T. Conti, 1988)

team really acting as a team, is essential. Otherwise, the opportunities for improvement will be limited to isolated in-process and intra-functional activities, and the improvement process will encounter barriers which will prevent it from being total.

The key word to remember when formulating an improvement strategy is 'incremental'. Continuous improvement is an incremental process. Attempts to make a big dramatic breakthrough all at once will not succeed because an organisation is too complex for a grandiose 'quick fix'. Having said this, sometimes quantum improvements are made. Everyone in the organisation has to learn from experience, and the soundest way of achieving the kind of cultural change which is needed and which lasts is to start simply, and progress in successive stages from individual team improvement projects which will have a demonstrable impact on the 'bottom-line' of the business and on managerial and employee confidence, to improving actual processes and streams of processes.

We repeat that companies should be prepared to start simple – everyone in the organisation from the senior management team downwards is at the bottom of an experience curve, and it is a safe bet that few people in the organisation will have experience of managing improvement projects – let alone cultural change, and holding the gains. Therefore, organisations must be prepared to invest heavily, in time as well as money, in a planned programme of education and training. Effective education and training does not necessarily have to be expensive in terms of cash, but it is expensive in terms of time – this should be considered as an investment rather than a cost.

## Standardisation

Once an improvement has been made, it needs to be incorporated into the relevant company standard and working practises, otherwise it can be lost and there might be a tendency to revert back to the status quo. The improvements will also identify areas in which standards need to be established. If standards on procedures, administration, system, component, product, etc., are not updated to reflect the improvements made, this could lead to organisational chaos, particularly in the field.

It is also important that the improvements made are transferred to other processes, sites, branches, etc., to reflect the good results of performance on a company-wide basis. An organisation will have to establish some company-wide mechanism or system to facilitate this and to keep standards up-to-date.

An investigation into standardisation will, in most organisations, reveal that many standards are not controlled or utilised, and do not meet the company requirements. Standards relate to procedures, rules and norms, and it is important that there are adequate procedural standards for each process – consequently, there needs to be a positive organisational commitment to standardisation.

The Kaizen strategy of continuous incremental improvement followed by standardisation of the gains, necessitates the development of a company-wide quality system in which the human, administrative and technical factors affecting quality are defined, clearly understood and managed. The goal is to ensure that every time a process is performed, the same information, methods, skills and controls are used in the same way in order to achieve a consistent output. The function of the quality system is two-fold:

- To underpin the integrity of the company's quality assurance responsibility so that it provides customers with products and services that consistently meet an agreed standard within the constraints of its existing process capability.
- To provide a reference framework on which to build systematic process improvement activities.

Implementing a quality system that meets the requirements of the internationally recognised standard ISO 9000 series[31] is a useful way of establishing such a reference framework. However, it must be stressed that this is a minimum or base requirement.

## Strategic dynamics of Total Quality Management

The overall business environment is not static, but subject to constant change. There are dynamic, motive forces which shape improvement strategies over time. We believe that there are six fundamental forces, which we have called the 'Strategic Dynamics' of TQM. These are:

- Customer care.
- Competitors.
- Leadership.
- Suppliers.
- People.
- Money.

They provide a basic framework on which to benchmark the progress an organisation is making towards TQM.

**Customer care**

Customer care is concerned with the nature of the customer relationship and managing the interactions between a company and its customers – what in the literature on service quality are called 'moments of truth'. The underlying assumption is that customers are not market segments, they are people – just like the rest of us – and need looking after too. Customer care means more than just reacting to customer wants and needs, it must be a pro-active process anticipating wants and needs, and with a planned approach to meeting them. This requires education and a managerial and communications infrastructure which is attuned to the customer, but not only to the external customer. As already mentioned, anyone within an organisation to whom a service is provided is a customer, so customer care must also have an internal focus. Unfortunately, many customer care programmes which are employed in service environments rarely focus on the production processes. The whole company must be centred on and driven by customer-oriented change.

**Competitors**

Business – the profitable interaction between supplier and customer – is conducted in a dynamic, competitive environment; therefore, competitive analysis is crucial. As we shall see later, this activity is often taken for granted, and sometimes forgotten. Moreover, competitors are a valuable source of information, ideas and inspiration for sustaining and enhancing TQI. Competitive benchmarking and comparing corporate performance against competitors provide an important basis for differentiating the company's strength in the market-place. Companies seeking to create competitive advantage should not limit benchmarking exercises to an internal organisational analysis or that of their competitors. Companies aiming to be the best must be prepared to learn from leading companies in other industries and emulate their best practises.

For example, two of the best-in-class processes emulated by Xerox (a 1989 Malcolm Baldrige Award Winner and a superior performing company) are the customer billing system of American Express and the warehouse distribution system of L. L. Bean (the US outdoor sporting goods retailer and mail-order house) – companies with core businesses which are very different from that of Xerox.

**Leadership**

Transforming leadership – that is, the chief executive acting as a value-shaper creating organisational purpose – is a crucial factor in TQI. It is not a once-and-for-all input by top management. Leadership which is about coping with change – as opposed to management which is about coping with complexity – is not limited to top management. Leadership opportunities abound within organisations undertaking the task of TQI. Top management must ensure that there are cadres of people at all levels within the organisation equipped to exploit these leadership opportunities. The leadership issue is discussed in Chapter 5.

**Suppliers**

Suppliers have a major impact on the organisation's ability to be competitive and to satisfy its customers. So for any organisation engaged in a process of continuous improvement, it is vital to seek improvements in the ability of its supplier base to consistently deliver conforming products and services, and to integrate suppliers into its improvement strategy. Technological and competitive pressures have resulted in a greater trend towards specialisation and, therefore, a higher degree of dependence on the expertise of suppliers. The implementation of an effective improvement strategy will depend in part on how well this supplier expertise has been deployed. This means treating suppliers as long-term and collaborative business partners, and not as adversaries. As with customer care, supplier relationships need careful nurturing, which means building close links based on trust.

Supply-chain management is discussed in Chapter 6.

**People**

People are the most important component of TQI. It has been claimed many times by world-class organisations that people are their most valuable resource. We believe that organisations are nothing more and nothing less than collections of people, and that successful organisations have a shared vision and sense of purpose. We believe that TQI begins and thrives with educated people who are provided with the means to apply their new skills to bring about change within an organisation – that is, 'empowerment'. It is important that organisations invest not only in technology and systems, but also in their employees' education.

An often quoted remark from commentators on Japanese management practices is that *quality begins and ends with education.*

Motorola (a 1988 Malcolm Baldrige Award Winner) has an annual training budget of $60 million, much of which is related to quality improvement. Its Employee Development Policy requires that each of Motorola's 100,000 employees world-wide undergoes a minimum of 40 hours of job-relevant training and education. The extensive Directory of Program and Services offered by Motorola's Training and Education Center has 28 days of material related to quality tools, strategy and techniques. Motorola expect all their engineers and manufacturing middle managers to have taken all 28 days of training during the course of the next few years.

**Money**

According to Juran,[6] money is the only language understood by upper management. Dale and Plunkett[34] also describe how senior management becomes switched on to the need for TQM when failure data is presented in quality cost terms. We are prepared to give senior managers the benefit of the doubt and assume that many are fluent in other languages too, and that they listen to and understand the voices of the customer and the supplier, and of their business processes. But money does provide proof of the need for action and concentrates the managerial mind as nothing else can. TQI is about increasing the added value of the organisation's process network, and added value can only be expressed in terms of money.

## Processes, organisation and culture defined

Throughout this book we refer to cultures, organisations and processes. Therefore it is important to explain what we mean by these three distinctive terms.

*Process* – The flow of work activities to discharge a specific operational requirement. Processes include all the functional and interfunctional activity flows that generate 'products' for external customers and for internal use. The term 'product' is used in the broadest sense to cover goods, services, information, documentation, software, etc. Processes are the building blocks of the organisation.

*Organisation* – The clearly defined and bounded management structure evolved for transacting business. The efficiency of an organisation is usually measured in terms of its ability to add value to inputs by managing various process transactions. In a passive sense, the organisation is a network of people, roles, responsibilities, processes and systems. However, an organisation is more than just structure and relationships; it is a group of people with a sense of collective purpose (see Chapter 3).

*Culture* – This is not so much what the organisation is, but more what its members think it is. Culture is the spirit of purpose, encapsulated by a set of values. But more than this, culture is a paradigm, a mind-set, a framework for understanding within an organisation. Basically, culture is the collective consciousness of an organisation.

The company/business/foundation/institution is the sum of its culture, organisation and transformation processes (as illustrated by Figure 1.16).

Processes are the basic building blocks of an organisation and, therefore, are a vital component of any quality improvement initiative. But just improving processes will not bring about a lasting change. Such improvements will almost inevitably be made in isolation, have a temporary effect and result in the cycle of decay and disappointment illustrated in Figure 1.10. Too many quality improvement ini-

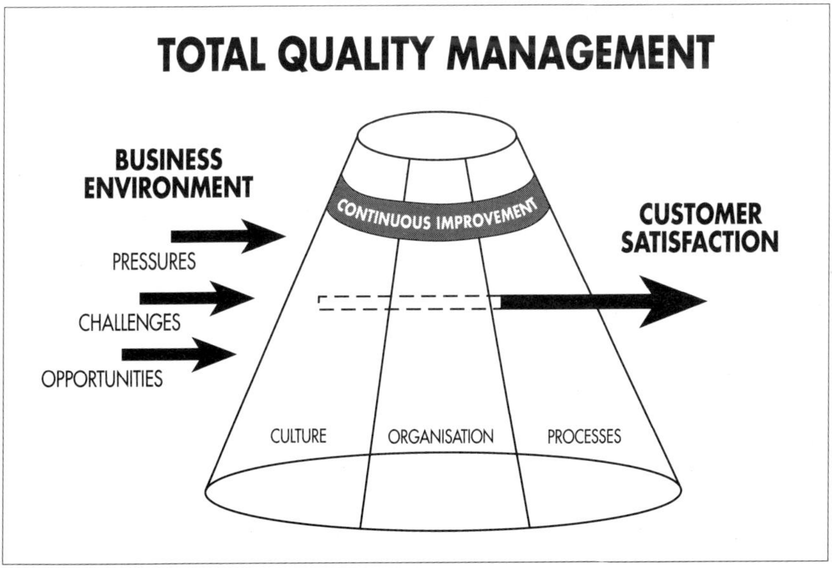


**Figure 1.16** Aligning the business to TQM

tiatives are limited to applying tools and techniques to improve processes (see Chapter 4). Organisations must change as well (there is more about this in Chapter 3). TQI necessitates a clear strategic direction for the whole of an organisation.

Like any business strategy, TQI requires careful planning and implementation by top management – this is the basis for driving organisational change. The long-term effectiveness and adoption of such change will be dependent on the cultural development of the organisation. Evolving a business culture based on Total Quality values, transforming leadership, teamwork, commitment to never-ending improvement and customer delight will take time – probably five to ten years. The transition from process improvement to organisational change to cultural change becomes progressively more difficult. In managing TQI, companies will need to address all three as early as possible.

The vision of any company seeking to attain superior competitive status should be to:

- Develop an organisation aligned to satisfying customer requirements 100% of the time, and in which everyone knows his or her role and has the skills and opportunities to fulfil it.
- Improve process streams continuously to meet and exceed changing customer requirements.
- Create a business culture in which Total Quality is a way of life, and where everyone in the organisation is committed to delighting customers.

Adaptation to the constant stream of pressures, challenges and opportunities from the modern business environment is perhaps the most important theme for business success. Survival in the current environment depends on achieving the goal of a totally well-managed business. Another 'motherhood' statement? We think not. To quote Deming :

*You don't have to do this. Survival is not compulsory.*[17]

## Summary

In this chapter we have presented the concept of quality as the manifestation of customer satisfaction, and developed the argument that quality management is a dynamic set of activities to achieve this goal. Quality management becomes truly total when it becomes an integrative framework encompassing the three strategic parameters of customer perception, competition and business performance.

The vision of TQM requires an enabling mechanism based on continuous improvement, and incorporating the strategic dynamics that drive the entire business organisation – we have called this Total Quality Improvement.

## References


1. Price, F., 1990, 'The Quality Concept and Objectives', in Lock, D. and Smith, D. J. (Eds), *Handbook of Quality Management*, Gower, Aldershot.
2. Houghton, J., 1990, Keynote speech given at the Second European Quality Management Forum, London.
3. Ford Motor Company, 1985, Three Day Statistical Process Control Course Notes, Ford Motor Company, Brentwood, Essex.
4. *Quality Vocabulary, International Terms*, 1987, BS 4778: Part 1, (ISO 8402 : Part 1), 1986, British Standards Institution, London.
5. Crosby, P. B., 1979, *Quality is Free*, McGraw-Hill, New York.
6. Juran J. M., 1988, *Quality Control Handbook*, McGraw-Hill, New York.
7. Parasuraman, A., Zeithaml, V. A. and Berry, L. L., 1985, "A Conceptual Model of Service Quality and its Implications for Future Research," *Journal of Marketing*, **49** (Fall), 41-50.
8. McConnell, J. D., 1968, "An Experimental Examination of the Price-Quality Relationship," *Journal of Business*, **41** (10), 439-444.
9. Jacoby, J., Olsen, J. C. and Haddock, R. A., 1973, "Price, Brand Name and Product Composition Characteristics as Determinants of Perceived Quality," *Journal of Applied Psychology*, **55** (6), 570-579.
10. Geistfield, L. V., 1982, "The Price-Quality Relationship Revisited" *Journal of Consumer Affairs*, **16** (Winter) 334-335.
11. Gerstner, E., 1985, "Do Higher Prices Signal Quality?" *Journal of Marketing Research*, **22** (5), 209-215.
12. Buzzell, R. D. and Gale, B. T., 1987, *The Profit Impact of Marketing Strategy Principles: Linking Strategy to Performance*, The Free Press, New York.
13. Phillips, L. W., Chang, D. R. and Buzzell, R. D., 1983, "Product Quality, Cost Position and Business Performance: A Test of Some Key Hypotheses," *Journal of Marketing*, **37** (Spring), 26-43.
14. Luchs, R., 1986, "Successful Businesses Compete on Quality Not Costs," *Long Range Planning*, **19** (1), 12-17.
15. Van Nievelt, M. C. A., 1989, "Quality, the Superior Index of Performance," *Proceedings of the European Quality Management Forum*, European Foundation for Quality Management, Eindhoven, 65-70.
16. Walsh, M., 1987, "Management of Quality as a Strategy," *International Journal of Quality and Reliability Management*, **4** (2), 38-52.
17. Deming, W. E., 1982, *Quality, Productivity and Competitive Position*, MIT Press, Cambridge, Massachusetts.

18. Handy, C. B., 1989, *The Age of Unreason*, Business Books, London.
19. Townsend, P., 1986, *Commit to Quality*, John Wiley.
20. Teboul, J., 1991, *Quality Dynamics*, Prentice-Hall, Hemel Hempstead.
21. Gummesson, E., 1987, *Quality – The Ericsson Approach*, Ericsson, Stockholm.
22. Zeithaml, V. A., Parasuraman, A. and Berry, L. L., 1990, *Delivering Service Quality: Balancing Customer Perceptions and Expectations*, Free Press, New York.
23. Feigenbaum, A. V., 1982, "Quality and Business Growth Today," *Quality Progress*, **15** (11), 22–25.
24. Deming, W. E., 1945, *Statistical Method from the Viewpoint of Quality Control*, Lancaster Press.
25. Juran, J. M., 1945, *Management of Inspection and Quality Control*, Harper and Row, New York.
26. Hill, T., 1985, *Manufacturing Strategy*, Macmillan, London.
27. Kotler, P., Fahey, L. and Jatusripitak, S., 1985, *The New Competition*, Prentice-Hall, Englewood Cliffs, NJ, USA.
28. Ishikawa, K., 1985, *What is Total Quality Control?: the Japanese Way*, Prentice-Hall, Englewood Cliffs, NJ, USA.
29. Feigenbaum, A. V., 1983, *Total Quality Control*, McGraw-Hill, New York.
30. Dale, B. G., Lascelles, D. M. and Plunkett, J. J., "The Process of Total Quality Management," in Dale, B. G. and Plunkett, J. J., 1990, *Managing Quality*, Philip Allan, Hertfordshire.
31. *Quality Systems*, 1987, BS 5750 (ISO 9000) series, British Standards Institution, London.
32. Imai, M., 1986, *Kaizen: The Key to Japanese Competitive Success*, McGraw-Hill, New York.
33. Conti, T., 1988, "From Processes to Streams of Processes: Overstepping the Borders of the Traditional Domain of Quality," *Proceedings of the 33rd European Organisation for Quality Conference*, Vienna.
34. Dale, B. G. and Plunkett, J. J., 1991, *Quality Costing*, Chapman and Hall, London.

# Chapter 2
# Total Quality Commitment: the Japanese Approach

## Introduction

Most experts are agreed that quality is the dominant factor in the success of Japanese companies in world markets, and much has been written on this during the last decade or so. In any text dealing with the subject of TQI, it would be a serious omission not to include some discussion of the way in which Japanese companies manage quality.

TQC is undoubtedly the integrative strategic framework of the Japanese company. The concept of TQC (which readily translates to Total Quality Commitment or Total Quality Care) is accepted without question by the Japanese, and it is the qualifying criterion in their home market. TQC is not perceived as desirable, but as essential for continued survival. Japanese companies through their considerable efforts over the last 25 to 30 years have put the principles of TQC firmly into place and are totally committed to sustaining the process of continuous improvement.

This chapter is structured under the following broad headings: customer satisfaction, long-term planning, research and development, organising and planning for quality assurance and improvement, management of improvement, visible management system, involvement of people, education and training, Total Productive Maintenance and Just-in-Time. The data on which the chapter is based has been collected from four study missions of Western European manufacturing executives to Japan; they examined the approach of a selection of their major manufacturing companies to TQC. Some comments on comparisons with European companies are made at appropriate points in the text.

## Customer satisfaction

In Japan, the internal market-place is dominant, and competition is fierce. This means that organisations need to be totally dedicated to satisfying customers. This effort must be long term and continuous, otherwise they will be overtaken by the competition. The market is saturated and demands:

- Ever-increasing product diversification and attractiveness.
- Speedy response to market needs.
- Rigorous reliability.
- Quality of conformance.

Japanese companies believe that bringing new products to the market-place quickly is the means by which they can sustain their competitive edge. 'The customer always comes first' is a term used by organisations to describe their market-oriented spirit. They are forever looking at the needs of the market. Japanese managers also comment that their customers' quality requirements are becoming increasingly rigorous, and that these requirements are a moving target. Take the example of the 'camcorder'.

At the start of the 1980s, the video camera/recorder was a large and bulky object weighing around 2.5 kilograms. It was inconvenient and used mainly at home. In the mid-eighties, Sony brought out a camcorder that was three times smaller and lighter. Suddenly the video camera/recorder was no larger or much more expensive than a good SLR camera. Almost overnight, Sony had widened the market to include people going on holiday. Sony's new product was so successful that it increased video camera sales in Japan from 1.3 million units in 1988 to 2 million units in 1990. It also produced a flurry of competitive offerings, each with new and improved features. For example, Matsushita included special electronics to cancel out the effects of shaky hands. Sanyo styled its offering like a pair of binoculars. Hitachi featured automatic focusing. Chinon's product weighed less than 700 grams. The net result was £3 billion of world-wide sales of camcorders by Japanese companies.

It is hardly surprising, therefore, that their most important measures of success are market-share and net sales, rather than rate of return on investment.

There is a total belief that business operations and efficiency can be improved by reflecting the customer's needs and requirements, and Japanese organisations have a variety of systems, procedures or mechanisms by which they can properly identify these needs and keep focused on the market. They go to considerable lengths to collect information on the wants and needs of customers. For example, a manufacturer of ceramic products has 4,500 fixed points of observation from which data is collected.

Japanese companies believe it is important for the engineers who are involved in the development of new products to go to the locations where equipment is being used and ask users, including field operators, questions such as:

- How do you feel about the current product?
- What are your main causes of concern?
- How does the product compare with that of the competition?
- What features should the new products have?
- What is required to satisfy customer needs, expectations, thinking and ideas?

This knowledge, together with that accumulated through other means of listening to customer voices, is used to pinpoint the technical gaps of the competition and identify product features which the customer finds attractive and charming, thereby helping to differentiate the product from that of the competition. It is also common practice for Japanese organisations to use Quality Function Deployment in conjunction with the seven new quality control tools, e.g., relational diagram, affinity diagram, systematic diagram, matrix diagram, matrix data analysis, process decision program chart, and arrow diagram. Mizuno[1] provides details for clarifying the required quality (objective and subjective), and translating customer wants into design requirements and build-in quality at source.

Detailed information is also developed on customer profiles, their current needs and future expectations. The data banks which Japanese organisations have built up on this are far in excess of anything witnessed in European companies.

## Long-term planning

### Quality, cost and delivery

Quality (including service), Cost, and Delivery (accuracy and lead time) are the main objectives, and organisations strive to become the best-in-class in relation to QCD. This is a prime consideration in company vision, mission and policy statements. The evidence collected indicates that this has been the case in most Japanese companies for the last 30 years or so. Extensive use is made of mottos expressing some appropriate message on QCD. This assists in keping the theme in the forefront of employees minds. Mottos, which are often suggested by employees, are changed at least once a year. It was stressed by representatives of the companies studied, that this policy of 'quality first' has remained unchanged since the formation of their organisation. They believe that their corporate strength is built up through TQC and they state that quality is foremost in every aspect of organisational corporate policy. This view is encapsulated in the point made by one organisation that even if only one out of 10,000 products failed, the failure rate for that customer would be 100%.

**Policy deployment**

Planning, feedback and decision-making for TQC are long term, often extending at least ten years into the future, and then a series of middle-range plans are formulated to assist in meeting the long-term business plan and strategic themes.

One of the main TQC planning activities is the deployment of the president's annual management policy plan (developed from the company's long-range and mid-term plans) to all levels of the organisational hierarchy. This plan is made available to the group companies at the beginning of the fiscal year. The deployment is carried out in the first place by the plant managers in their respective manufacturing divisions, and then the plant manager's policy is successively deployed by each section/department manager in their area of responsibility, through to foreman and line operators. The deployment is in terms of QCD. Figure 2.1 is an illustration of the policy deployment system from a manufacturer of ceramic products.

Each plant manager develops his/her annual policies and improvement targets for every section and department of the plant(s) which are within his/her remit of responsibility to meet the president's policy. He/she decides the annual policy for the plant, and what key problems need to be tackled in relation to the president's policy. The target is based on the long-range business plan, long-range plan for the plant's operation, and improvements which need to be made, taking into account an evaluation of the year's activities, production schedules, etc., from the previous year's performance. This target is fully discussed and debated with each section manager in relation to their annual policies and plans for departmental activities, until a final target is agreed through consensus, along with the methods to reach the goal.

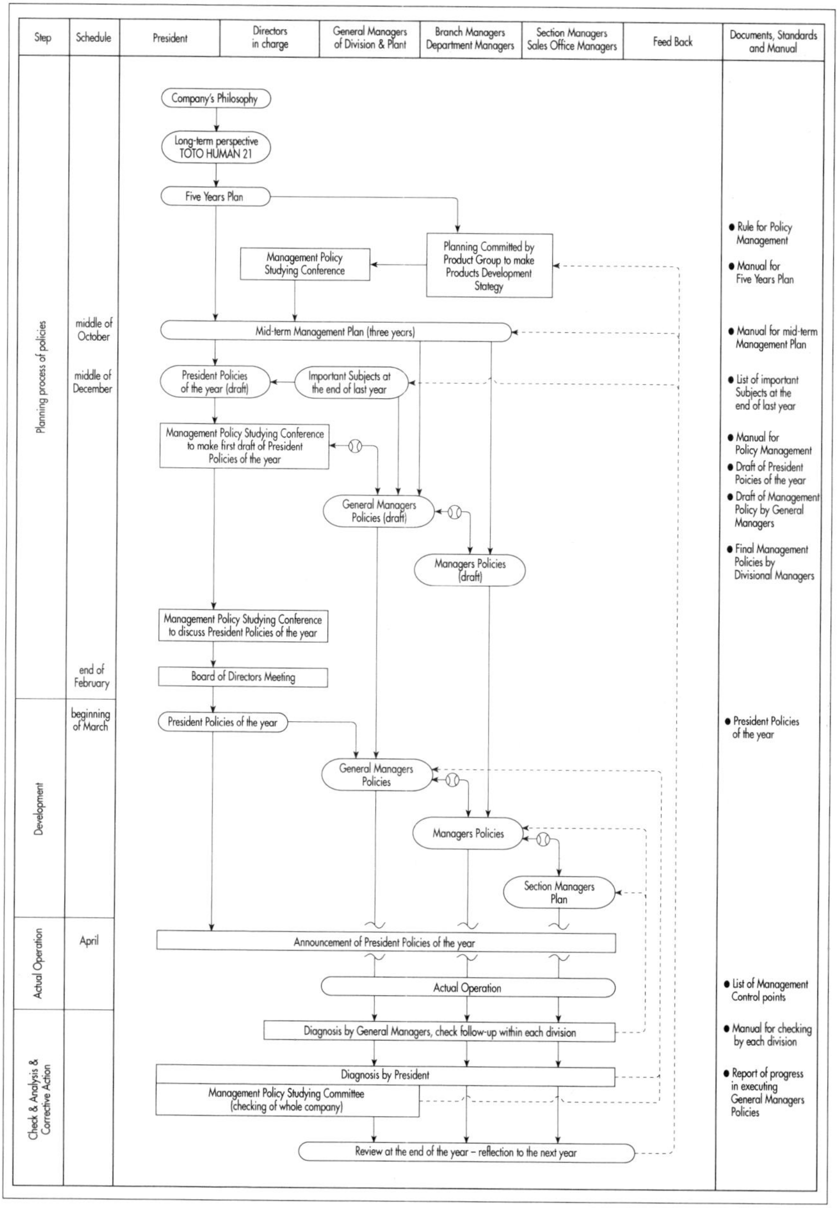


**Figure 2.1** An example of the policy management system (Source: Toto Ltd., Chigasaki Works, Chigasaki City, Japan)

For example, the plant manager may suggest a target of 5% improvement and the section manager may reply with a suggestion of 2% – this process continues until a target is agreed with which all parties agree is feasible within the time frame of the plan. This particular activity is called 'play/catch'. It is usual to set yearly and half-yearly improvement plans and targets. The section manager then agrees with each of his/her foremen the activities, plans and targets for their group, who in turn agrees roles, targets and improvement activities with each operator. Each division keeps a register of the improvement action agreed with staff.

There is a set time-scale (usually six to eight weeks) for this policy management deployment activity to be cascaded down through all organisational levels. In each company, the policy deployment commences at a set time in the calendar year.

The control of policies is generally made through:

- The president's diagnosis.
- Plant manager and section manager diagnosis and monthly review of plant activities.
- Discussion of achievements and improvements at plant conferences on QCD.
- Each section's daily management and records in terms of clarity of the section's function in the organisation and the role of each person, detailing clear points of control, activities for improvement, standardisation and taking corrective action.

The Plan, Do, Check, Action cycle is extensively used in all these diagnoses.

It is usual for the plant manager to audit on a quarterly basis the progress being made by each section to achieve its improvement objectives, and for the section manager to undertake a diagnosis of his section at quarterly and monthly intervals. It is also usual for the line operators to carry out a self-estimation of their achievements, against the agreed target, and this written assessment on a scale of 1 to 5 is commented upon by the foreman and followed up by a personal interview and, where appropriate, methods and plans are revised. The foreman then reports to the section manager on his achievement, the outstanding problems and the priority actions to be taken, and this reporting procedure is continued up through the organisational structure. Typically, each person/group completes each month a report of their improvement projects and activities which have been agreed, and the progress being made against the plan. This also applies to management, and each manager is expected to register their own improvement projects which are linked to those of the manager/director to whom they respond. If there is low achievement against the target, a full and frank discussion will take place with all concerned to determine the reasons for this, and to decide upon corrective actions to recover the situation.

In some organisations, the process of policy management deployment is also subject to diagnosis by outside experts such as JUSE and/or university professors. Any reports relating to the diagnosis are subjected to an in-depth examination. It is argued that it is the process of the deployment which is important, and not just the results.

It is usual for each section to have a visible display of this policy deployment as part of their visible management system. Figure 2.2 illustrates the key points of the typical format of such a display. The left hand side of the chart shows the tree of policy deployment from

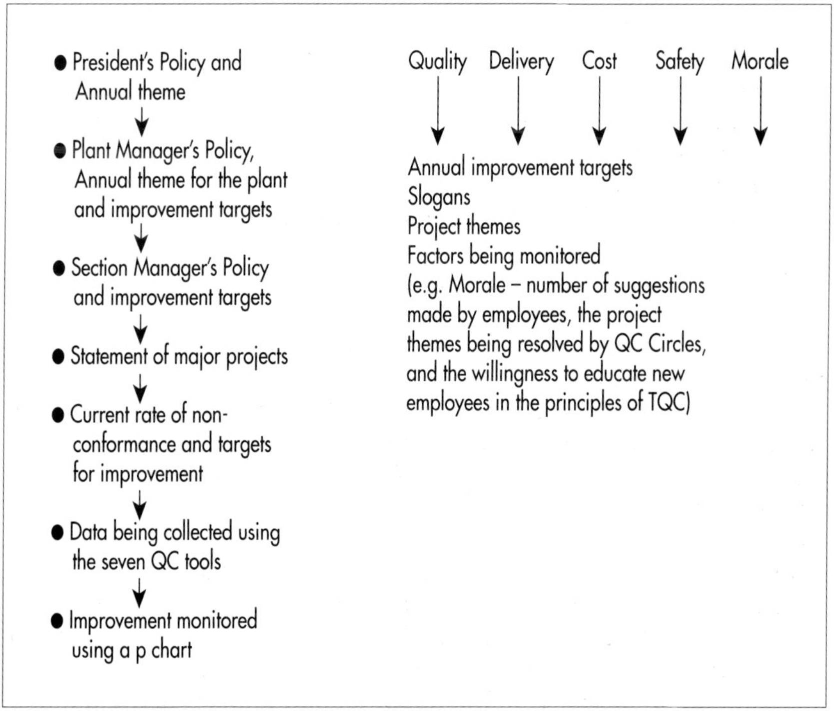


**Figure 2.2** Visual display of policy deployment

the plant manager down to the level of each section; each section makes their own plans for improvement based on existing problems. The overall rate of imperfection for each section is related to the different processes, with information being provided on individual problems. A proportion defective (p) control chart is used to monitor the rate of imperfection against the set target. The right hand side of the chart displays annual improvement targets for quality, cost, delivery, safety and morale. A slogan relating to the improvement is displayed on the board adjacent to the problem to be solved. The names of the workers who are responsible for the various activities relating to the policy deployment are also displayed.

All these plans include a number of provisions for the involvement of employees in the quality improvement process. Positioned at right angles to the board and completing the policy deployment corner are pictures of typical imperfections in the section, together with an improvement book. This book logs each improvement that has been made, and helps to promote standardisation, serving as a point of reference for people on what type of improvements have been made in the past.

Within each section there is an effort to solve at least two major projects each year, these projects derived from the policy deployment having been registered as themes. This is in addition to the activities of QC circles. In European companies there is a tendency by departments to try and solve too many problems at the same time, and consequently the improvement effort tends to be too thinly spread.

In European companies the job of chief executive officer is a lonely one and it is often the case that the respective managing directors, plant managers and other middle managers filter out information to the chief executive officer. In Japan, the policy deployment and its visible display at shopfloor level not only ensures that the president has a window on what is happening in each section of each division under his control, but that every person is clear on the details of the president's policy and what is required by their section and themselves to make the necessary contribution to the achievement of overall policy. The discipline of policy deployment, and the agreement at each organisational level of achievable targets, ensure that the energy of line operators is directed in the same direction as section managers, and so-on, up to the level of the president – this is often termed a 'spiral upward movement'.

There is little doubt that the policy deployment method facilitates the attainment of corporate goals, and facilitates the operation of the organisation in a systematic manner. It also assists in integrating the quality improvement process with organisational long-term strategic plans.

**The role of senior management**

The total commitment and leadership of senior management on a long-term basis is always stressed by the Japanese as being a key point for successful TQC activities. The role of senior management includes:

- Ensuring that the entire organisation is committed to TQC, and establishing corporate quality systems.
- Promoting TQC activities continuously.
- Participating in activities such as:
  - membership of the committee for quality planning.
  - quality assurance meetings for design and manufacture of quality into the product.
  - quality audit, quality improvement and corrective action meetings.
  - quality audits and diagnosis of the improvement activities of site locations. In relation to these audits and diagnoses, senior management study, prior to the actual visits, information on factors such as the quality improvement plans, targets, achievements and problems (this data is provided by the TQC Promotions office).
- Motivating employees.
- Involvement in quality-related education and training.

## Research and development

Japanese companies believe that Research and Development (R&D) is the main way to sustain their competitive edge, and this is the major focus of their efforts. They invest heavily in R&D and vigorously pursue new product development. Japanese companies are engaged in short-term and long-term R&D. The long-term R&D relates to materials development, and how to combine and integrate different types of technologies. The less innovative short-term R&D is geared to the development of new product features. They tend to engage in pure research as opposed to leaving this to universities.

The aim of Japanese companies is to bring an increasing number of attractive high-value added products to the market in the shortest possible time. This is seen by them to be a key issue, and is one to which they commit considerable time and resources. Product life-cycles in Japanese markets are becoming shorter and shorter, and there is an ever increasing demand from the market-place for new products. Considerable effort is concentrated on reducing the cycle time for product development. The Japanese market is forever looking for products with more unique features. Japanese companies believe that if they produce standard products with only a minimum of diversification they will not survive in the market-place. Some Japanese companies have combined the R&D and marketing functions to facilitate the creation of new market demands, develop their present market-share and exploit technological know-how.

Japanese companies have a systematic approach to R&D, and this helps to reduce the time between the conception of an idea and the market launch. They have a huge R&D database on their accumulated experiences – from product development to full-scale production on which their designers can draw in order to satisfy customer needs. This typically contains a variety of information on aspects such as:

- Design features.
- QFD carried out on previous products.
- Product development.
- Design for manufacturability.
- FMEA.
- Fault Tree Analysis (FTA).
- Success Tree Analysis (STA).
- Reliability.
- Product shapes and features which appeal to customers.

This is the result of many years accumulated experience, and there is little doubt that this resource provides a competitive advantage by enabling them to produce new designs and products at a faster and faster rate; one which European companies will find hard to emulate. Japanese companies thrive on their reputation for leading-edge technology, and corporate prestige is measured in terms of the R&D activity.

The R&D project team steers the product concept through the various stages from R&D to development through to trial production and ultimately full-scale production. Therefore the R&D team ensures that the design intent is being planned into the product, helps the production division understand the product, assists with operator training, and deals with problems as and when they arise during each stage of the manufacturing process. A commonly expressed view is that the Research Laboratory is the teacher and the factories are the students. The R&D Centres also tend to select and prove the equipment before it is used by the factory in full-scale production.

At one Research Centre it was clear that, as suppliers of components to the automotive industry, they were developing advanced electronics for use in automotive components, and that in turn these developments were driving design developments in the major motor manufacturers. In general, the reverse is true in the European motor industry, where the supplier manufactures to an original equipment manufacturer's design, and conducts little development work of his/her own.

## Organising and planning for quality

### Promoting Total Quality Control

Japanese companies take a company-wide approach to the assurance of quality from product planning through to sales and services. They typically have a TQC promotions office at their head office and at each plant. It is used to promote TQC through a variety of activities such as:

- Establishing a TQC policy.
- Education and training.
- Promoting standardisation.
- Cross-functional teams and steering committees.
- Analysing and co-ordinating improvement activities.

### Cross-functional committees

The Japanese are great advocates of the lateral management of major functions and typically have committees dealing with:

- Quality assurance.
- Development.
- Cost.
- Delivery.
- Supply.
- Policy.
- Standardisation.

For example, a typical quality improvement committee will meet three or four times each year, and will establish improvement policies and deal with issues such as the organisational activities of QC circles and how to develop the skills of employees. A quality assurance committee will analyse, catalogue and discuss any day-to-day problems or non-conformance problems in the field, and make decisions on how to resolve problems.

**Quality-related responsibilities and roles**

The manufacturing department is responsible for maintaining quality. The quality assurance department is responsible for:

- Providing guidance to manufacturing and other sections in terms of problem analysis and developing improvement plans.
- Evaluation of product quality performance.
- Audits of the manufacturing division.
- Product inspection.
- Quality-related training.
- Ensuring that people follow up the plans which have been decided by the TQC promotion office or TQC committee, and assist in cascading these plans down to all levels of the organisation's hierarchy.

In large organisations there is usually a corporate quality assurance department (CQAD). The role of the CQAD at a major manufacturer of electronic products is to:

- Give guidance on TQC to all companies within the group.
- Set long-term and medium-term policies for the manufacturing divisions.
- Provide quality education and training.
- Provide quality auditing.
- Maintain a quality performance system for each division and to grade divisions according to performance.
- Assess whether the product is easy to use.
- Determine if the product is safe if it is used incorrectly.
- Carry out inspection of products on a component by component basis.
- Examine packaging.
- Carry out endurance tests.

- Undertake comparative studies and evaluation of their own and competitors' products, and other tests, to anticipate problems before they occur.
- Undertake life style research.
- Study how to produce readable instruction manuals.

In order to reach a consensus on the promotion of TQC activities and increase the level of corporate awareness of TQC and exchange quality information, regular meetings between corporate and plant quality assurance departments are held.

The general manager of each division is responsible for TQC, and the TQC promotion office and quality assurance department work together to facilitate continuous quality improvement. Each section and division is responsible for QCD planning at source, and submits a report on their improvement activities to the quality assurance department. The department then compiles a report for the TQC promotion office, which is in turn passed to the president for consideration in his annual audit of the division. This report typically covers:

- Quality improvements.
- Goals and achievements.
- Explanation of achievements and/or shortfalls.
- Projects undertaken.
- Feedback data.
- State of health of cross-functional teams.
- Customer satisfaction.

## Quality assurance procedures and systems

The Japanese hold the view that quality assurance is the central core of TQC and that TQC is difficult without effective quality assurance procedures. It was mentioned to us on a number of occasions that quality assurance is part of the Japanese style of thinking. Consequently, intensive effort is devoted to the assurance of quality on a day-to-day basis. All the host companies had charts and diagrams which outlined, in considerable detail, their basic quality assurance system and procedures.

When problems occur they are analysed in detail using a defect analysis sheet, and everything is itemised in considerable detail. The Japanese place considerable emphasis on finding out where and why they are doing things wrong. The usual procedure is to put in place some emergency measure, followed by individual and then systematic reoccurrence prevention measures. They are very careful not to repeat any failures made in the past, and make continual reference to various failure recurrence prevention check-lists. The Japanese are fanatical about taking precautions with regard to quality assurance, and when anything unsatisfactory is detected it is fed back to the appropriate upstream stage and preventative actions are taken to counteract the trouble. The emphasis is on source control. For example, in a transformer manufacturer they emphasise the theme, if you do not observe a such and such factor, mistakes will happen (with graphical details of the mistakes being displayed). Other typical activities employed for source control include:

- Quality assurance tables, e.g., control plans, for in-company and sub-contractors work.
- Design review to prevent any failure on the part of designers.
- Production of operating procedures by the foreman and line operators.
- Standards and instructions for daily control.

A variety of aids are provided to ensure that operators have all the help required to get it right first time and prevent errors from occurring. These include:

- Check sheets.
- Operating instructions.
- Product identification cards.
- Mistake-proof devices.
- Process operation sheets.
- Features and parameters to which attention is required.
- Machine vision systems.

At a furniture manufacturer, to promote process management and control at source, every line worker produces a working instruction entitled 'What I know about my job – the knack of doing my job'; these are displayed at appropriate prominent places.

In planning for quality, considerable attention is given to listening to the voice of the customer right from the R&D and design stages. In the stages of planning and production preparation, trial production and full-scale production and from field experience, detailed notes are kept of any problems encountered and the countermeasures which have been put into place. Taking notes is part of the Japanese management style! These notes are always referred to in the planning of new products. All the necessary preparations are made in advance of actual production, and considerable resources are committed to this activity. In European organisations, the production preparation stage is usually rushed, in the hope that any problems can be sorted out later. Even if notes are made, they are often not analysed and used to prevent reoccurrence of problems in future products, therefore there is no learning from past mistakes.

In the development and design stage, engineers from the quality assurance and inspection department take up residence in the development and design department (this is termed the Resident Engineer System). The designs are evaluated for potential difficulties at the volume production stage. Consideration is given to the preventative measures used for troubles experienced in the past, and efficient means of production to ensure design quality. In the production preparation stage, engineers from the development and design department take up residence in the production and inspection departments. This is to ensure design intent to previous processes, to feed in know-how obtained during the design and development stage, and to promote the implementation of countermeasures against troubles, including mistake-proofing.[2] These cross-functional teams facilitate the process of simultaneous/concurrent engineering, which not only reduces product development times, but also ensures that the designs are suitable for manufacturability and reduces the number of late engineering changes.

Major suppliers also join in at the design stage (they are called *guest designers*). This activity helps to ensure that the specialist expertise of the supplier is used, and that improvements are identified early in the design cycle. Suppliers are also encouraged to suggest cost levels for their products.

To help assure product quality there is feedback and/or feed forward of quality information in:

- Production planning.
- Product design.
- Evaluation of prototype products.
- Pre-production planning.
- Purchasing.
- Quality audit.

- Evaluation of pre-production products.
- Volume production.
- Inspection.
- Evaluation of the products from volume production.
- Sale and field service operations.

The objective is to build-in quality at each step before sending work to the next process. There is total collaboration and co-operation between R&D, technical, quality assurance and manufacturing departments to eliminate problems and ensure that processes are mistake-proof; this is facilitated through the use of cross-functional teams.

In order to identify defects early on in the design process and to assure the quality of design, it is usual to employ techniques such as:

- Design reviews.
- Design of experiments.
- Quality assurance meetings.
- QFD.
- FMEA.
- FTA.
- Quality audits.
- Reliability tests.

In the production preparation phase, the production engineers endeavour to predict failures for the process, and to take collective action before machine and process sequences are finalised. FMEA and process capability studies are employed to assist with this. It is usual to carry out a process capability study every time new production facilities are used, when a new design is produced on existing facilities, and in the mass production of established products.

In all the companies visited, housekeeping was immaculate. For example, a visit around a steel mill involved the mission members wearing white gloves, which when discarded were hardly discoloured. When touring the photo-setting department of a printing company, members had to tie plastic covers over their shoes. Little dust was found on window ledges in any of the organisations visited. Japanese companies believe in having clear gangways. Any necessary equipment relating to processes is stored in racks located on an outside wall. It is not uncommon to see the space between the gangway and the outside wall painted green to represent grass, and potted plants placed on the painted area. The discipline of cleanliness and housekeeping is a prerequisite for effective quality assurance. This should be pursued more vigorously by European companies. The general impression is of a working environment which is clean and comfortable and in harmony with employees.

## Management of improvement

It is crystal clear from the presentations made by the Japanese managers and technical specialists that they form part of a totally committed management team which is enthusiastic, enjoys work and is vigorously working on continuous improvement in pursuit of perfection. They clearly believe in the future and have a long-term vision of where their respective companies are heading. The Japanese companies visited exhibited the typical profile of high growth. For example, one company during the last three years had experienced an increase in sales of 21%, the new products to sales ratio having increased by 25%, and labour productivity by 50%. Most companies were planning to increase sales by 20-25% over a period of three years.

Japanese management teams articulate very effectively what they are doing, and have considerable confidence that the strategy and course of action they are pursuing is right. The aim of each company is to be the market-leader and best-in-class, whereas in the West the majority of companies appear satisfied with second or third place. Japanese companies tend to have a single title, or theme, e.g., Total Productive Maintenance (TPM), TQC, Just-in-Time (JIT), under which various initiatives are brought together, and this is translated into a company-wide effort. This helps to give their improvement activities and teams a clear focus. In European companies the initiatives being pursued tend to be somewhat fragmented, and are the responsibility of individual departments and people.

The Japanese have developed an organisational culture and management style which, based on the evidence of their investments in world-wide production facilities, can operate successfully anywhere. Their success is not simply a matter of national culture – this is but one factor. The key to their success lies in their ability to create an organisational culture conducive to continuous improvement. The saturation of Japanese markets, the strength of the yen, high labour costs and the aggressive sales drive of individual companies with expectations of higher and higher sales cause Japanese companies to seek an increasing number of off-shore manufacturing bases. A number of the study mission members expressed the view that "the Japanese could take over entire cultures."

An uncomfortable number of European manufacturing companies have a fundamental management problem, which is characterised by managers giving the impression that they are just going through the motions. In such companies, the improvement process will first stagnate and then finally degenerate if a small number of key people leave, whereas the Japanese appear to have moved to a situation of autonomous improvement. All employees manage the improvement effort for themselves, and it proceeds in a common direction with each person accepting ownership. The Japanese appear to have developed a standardised method of managing companies and improvement processes, a method that can be applied to most cultures. They are expanding on a world-wide basis in order to maintain their high sales-growth, and most companies visited have made a series of moves to internationalise their base of operation.

Through their involvement with individual initiatives, QC Circles and/or through the suggestion scheme, operators are constantly engaged in problem prevention and improvement activities. They also pay keen attention to the quality of machinery through Total Productive Maintenance. When problems arise, emergency teams assist the operator in rectifying the situation. There are regular meetings between groups of operators, supervisors and technical specialists to discuss problems and improvement actions. In European companies, when problems occur there are usually a number of organisational layers between spotting the problem, its solution and recovery of the situation.

It is worth pointing out that Japanese manufacturing managers operate in a more favourable environment than their European counterparts. Japanese companies have a bias towards manufacturing in general and engineering in particular. Companies employ huge numbers of engineers. The majority of European manufacturing companies employ an insufficient number of engineers and consequently have not the resources to solve problems. Every Japanese person with whom discussions were held during the course of the study missions appeared to believe that manufacturing is the key to their national economy. Improving manufacturing efficiency and productivity was discussed all the time. Work occupies the collective consciousness of the Japanese people, and they all realise that their future depends on it.

Japanese managers and technical specialists exhibit a caring attitude towards production and what is happening on the factory shop floor. Senior and middle managers regularly visit the shopfloor to see what is happening, asking about results and problems, and giving advice and help to create good habits through leadership (Management By Walking About). In European companies, most senior managers tend to isolate themselves in their offices.

The Japanese invest in equipment without having to worry about short-term payback periods because they know that this will be beneficial over the longer term.

This willingness to invest must be a considerable feature in the motivation of their managers and engineers. In the main, investment is to reduce labour costs. Over the last thirty or so years the Japanese have obtained considerable proof of the wisdom of this policy. The typical payback period for equipment is three to five years, compared to the one year which is typical in European companies; interest rates for borrowing money are also lower. In European companies the production director is also required to specify a break-even production volume. In relation to investment in new equipment, it is a common requirement for European managers to make the justification over a three shift operation.

European manufacturing managers frequently complain that their engineers always want computer equipment and software for the projects which they are undertaking. By contrast, whilst they make extensive use of computer-aided design and manufacturing systems, Japanese engineers concentrate a considerable amount of their improvement effort on doing the simple things well. They also spend more time on the shopfloor than their European counterparts. A considerable amount of the equipment employed by Japanese companies is relatively basic. The key factor is not the equipment itself, but how it is used to improve manufacturing efficiency; it is not simply a matter of investing in new technology. All the Japanese companies visited had their own internal machinery manufacturing division, and a vast amount of the equipment seen during the factory tours was customised. The proprietary equipment is often employed to eliminate waste and transportation between processes, and to facilitate good internal logistics.

The Japanese are never shy in sharing the results of their improvement activities with Western visitors, for example:

### A semi-conductor manufacturer

| Goals | Achievements in Three Years |
|---|---|
| Equipment | 85% utilisation |
| Manufacturing cost | 50% reduction |
| Failure rate | 10% reduction |
| Production | Two-fold improvement |
| Suggestions | Four per person each month |
| Accidents | Zero |

### A temperature controller manufacturer

| Function | Before Just-in-Time, Upgrade, Manufacturing Process (JUMP) | After JUMP |
|---|---|---|
| Production lead time | 3 days | 2 days |
| Inventory | | |
| – parts | 30 days | 3 days |
| – WIP | 12 days | 8 days |
| – finished product | 5 days | 2 days |
| Workforce | 13 | 6 |
| Space | $20m^2$ | $12m^2$ |

## Visible management systems

Japanese companies place considerable emphasis on ensuring their operating data is visible on the factory shopfloor. They believe that everyone in the company benefits from an open information system. A complete range of information is displayed, in a variety of formats. This data assists managers, technical specialists and operators to manage their processes more effectively, facilitates the process of continuous incremental improvement and identifies and publicises the improvements. It is also a common communication mechanism to keep employees in touch with what is happening, and provides a focus to help concentrate efforts, indicating to people when events are not going to plan and providing warning signals of all kinds of different events. Display devices are often set up by operators and first-line supervisors. In some cases, a display is related to a specific manufacturing section, e.g., who is responsible for specific activities, TPM achievements, QC circle members and projects, and a skill matrix including pictures of operators, and in others it is related to a particular topic, e.g., policy deployment, mistake proofing, QC circles, educational and safety. The following are examples of this visible management system:

- A display of the activities of QC circles in terms of where they are located in the factory, current projects, achievements and membership.
- One company, as part of a campaign to improve the plant safety performance, displayed on a notice board the safety actions to which each operator had committed themselves together with their pictures.
- In a software engineering department, a board displayed against each work station the name of the software engineer and the job on which he/she was currently working.
- In a metal-part punching department, the process instructions of jobs which were scheduled to be started at various times during the day were hung on a rack against a time scale from 0830 to 1930.
- A schematic layout in the sub-assembly and assembly areas of one organisation indicated the flow of work and, using different colours, the zone position of different types of operators and their role. The following system was used:

  *red* – full-time employees.
  *brown* – casual employees.
  *blue* – operators responsible for the supply of parts.
  *yellow* – employees who come to assist a section which has fallen behind its scheduled production target.

- The main steps of a JIT in terms of its purpose, targets, activities, and achieved improvements were displayed on a notice board.
- A complete wall of one plant was covered in charts relating to a variety of issues, including:

  – Details and pictures of improvements which had been made.
  – Safety achievements and performance.
  – Attendance statistics.

- Quality issues.
- Suggestions.
- Production targets.
- QC Circle activities.
- TPM activities.

- Another company hung on racks at the side of machines a series of large cards which operators had produced in order to train their peers on specific aspects of TPM.
- In a refrigeration unit assembly line, quality control check-sheets were displayed at each work station indicating the product characteristics which needed attention, the important processes in terms of quality, and the self-inspections and tests to be carried out.
- In one organisation, a number of sub-assembly areas were supplying assembly lines, across a gangway. Each sub-assembly area and line had scheduled production targets for specific times of the shift along with actual production achieved. If production was going to schedule, a grey card was displayed across the gangway, whereas if production was two units behind schedule, a red card was used.

There is little doubt that this sharing and diffusion of information helps to ensure that everybody is working together for the good of the company, and will inevitably reduce organisational conflict. There must be a total openness about concepts such as TQC, TPM and JIT. Everybody in the company needs to know why they are doing it, the strategies and objectives, techniques employed, and successes and failures: this facilitates team spirit. In a number of companies the information displayed indicated that the improvement objective was to reduce the number of operating staff in a production section. Any staff affected in this way were transferred to other sections and to subsidiaries. For example, in one company operators were sometimes transferred to the sales function.

The pump-manufacturing division of a mechanical engineering company suffers considerable seasonal fluctuation with some 65% of production output scheduled between October to March. Middle and first-line management, together with the company labour union, make plans on how to adjust the workload of people in the manufacturing, assembly and engineering areas in both slack and busy periods. A matrix showing the movement and activities of people in these periods is communicated and displayed. For example, during the slack period, employees are moved between manufacturing, assembly and engineering to improve their skills and knowledge; maintenance, mistake proofing, standardisation, and improvement activities are carried out, jigs are produced, trial manufacture is conducted on parts currently purchased, and employees go to customers and dealers to assist with the selling of the product, listening to what they have to say about the product or service.

These last two examples emphasise the flexibility of the Japanese worker. All the companies visited stressed this point, and said that workers accepted job rotation and movement from one job to another very readily. This job flexibility and job rotation helps to eliminate the departmental boundaries which are often a major stumbling block to improvement activities; it also diffuses new technologies, approaches and systems to every corner of an organisation. Most workers are in multi-skilled groups and able to do all jobs in these areas; there are no detailed job descriptions, and the salary is paid for the person, not the job. Japanese companies firmly believe in developing, not specialists but good all-rounders. The system of lifetime employment and single company labour unions obviously facilitates flexibility, job rotation and long-term education and training programmes.

## Involvement of people

The mission statement of all the organisations visited stressed *that their greatest asset is manpower*, and all employees are encouraged to participate in quality improvement. In more than one company it was said that, because of the lifetime employment situation, they were always searching for ways to motivate and revitalise their staff. The usual means was through QC circles, suggestion schemes and other small group activities, as well as a variety of presentations, job rotation, and continuous education and training.

QC circles and suggestion schemes provide the mechanisms and motivation for involving everyone in an organisation in continuous improvement. Considerable importance is attached to these two activities, and they are viewed as complementary. A variety of recognition and award schemes, contests and prizes are in place for recognising employees' efforts, and providing direct rewards. Great kudos is attached to these awards. The quantity and quality of suggestions are seen to be important, and reflect the department manager's ability to create an environment that is conducive to improvement.

There is some difference in the way that suggestion schemes operate but, in general, the procedure now described is typical.

A suggestion is submitted and evaluated the same day by the foreman of the section, and the person making the suggestion gets a token sum of 100-300 Yen. If the suggestion is considered to have some potential it is passed to a committee (TQC Promotion or QC circle) which evaluates how many points it is worth, or alternatively the section manager can request more data from the proposer or QC circle, and this is evaluated in the same way. Formal acceptance and payment generally follow within one month of a suggestion being presented This fast reaction assists in maintaining the interest of employees. If considered of special merit, the suggestion can be put forward for monthly, six-monthly and annual commendations and the President's Prize of 10,000-20,000 Yen. This involves a presentation by the proposer to explain his/her idea, and an evaluation being carried out by the TQC Promotion and QC Circle Committee. There are a variety of methods for rewarding good suggestions.

Without exception, all the organisations visited had thriving QC circle programmes involving a large proportion of the total workforce, and suggestion schemes through which a considerable number of suggestions were obtained, most of which were implemented. QC circles can be considered as a natural part of Japanese working life, and in this respect it is perhaps not advisable to make a close comparison with the use of circles by Western organisations. Within each section, the Japanese line workers are involved with their foreman in making day-to-day improvements in their routine work activities. They do not appear to separate out, as is the case in the West, QC circle activities and day-to-day work activities; it is one unified improvement effort at shopfloor level. QC circles generally tackle projects which are related to their section's improvement objectives for meeting the president's, plant manager's, and section manager's policies, and they tend to be managed in this way. The organisations did stress that QC circles, whilst essential, are but a small part of their quality improvement process.

In most Japanese companies, all non-managerial employees belong to a QC circle. There is some suggestion that membership in manufacturing areas is not voluntary, but that considerable peer pressure is involved. The effectiveness of QC circles is usually assessed at annual or six-monthly intervals, at which time good performance is recognised.

Typical objectives of QC circles and suggestion schemes in the organisations visited are as follows:

### QC circles

- To provide opportunities for self-improvement of knowledge and skills through co-operative team efforts.
- To create a rewarding work environment.
- To create a workplace where total participation in quality control is a reality.

### Suggestion schemes

- To improve the power to work and individual abilities.
- To promote friendly and healthy human relations amongst all employees and to vitalise activities.
- To improve the company's structure and operations.

There is a considerable effort to create a work environment in which staff are active participants. The Japanese companies visited treated their workers as human beings and not as tools, and they went to considerable lengths to promote harmony between personnel and technology. A number of organisations said they would not lay their employees off in the case of a downturn in the market or due to a short-term fluctuation in demand, but would instead employ people in clearing and housekeeping activities. This not only assists the improvement process but facilitates the creation of a strong bond between employees and the company.

## Education and training

Japanese companies believe that everyone in an organisation must understand TQC and that this can only be achieved by education and training. They take a long-range view of quality education and training, and tend to have a master training schedule and curriculum to develop the skills of all their employees. The schedule recognises the different training requirements of people in different functions and levels in the organisational hierarchy (Figure 2.3) – basic courses (all employees), special courses (selected employees) and manage-

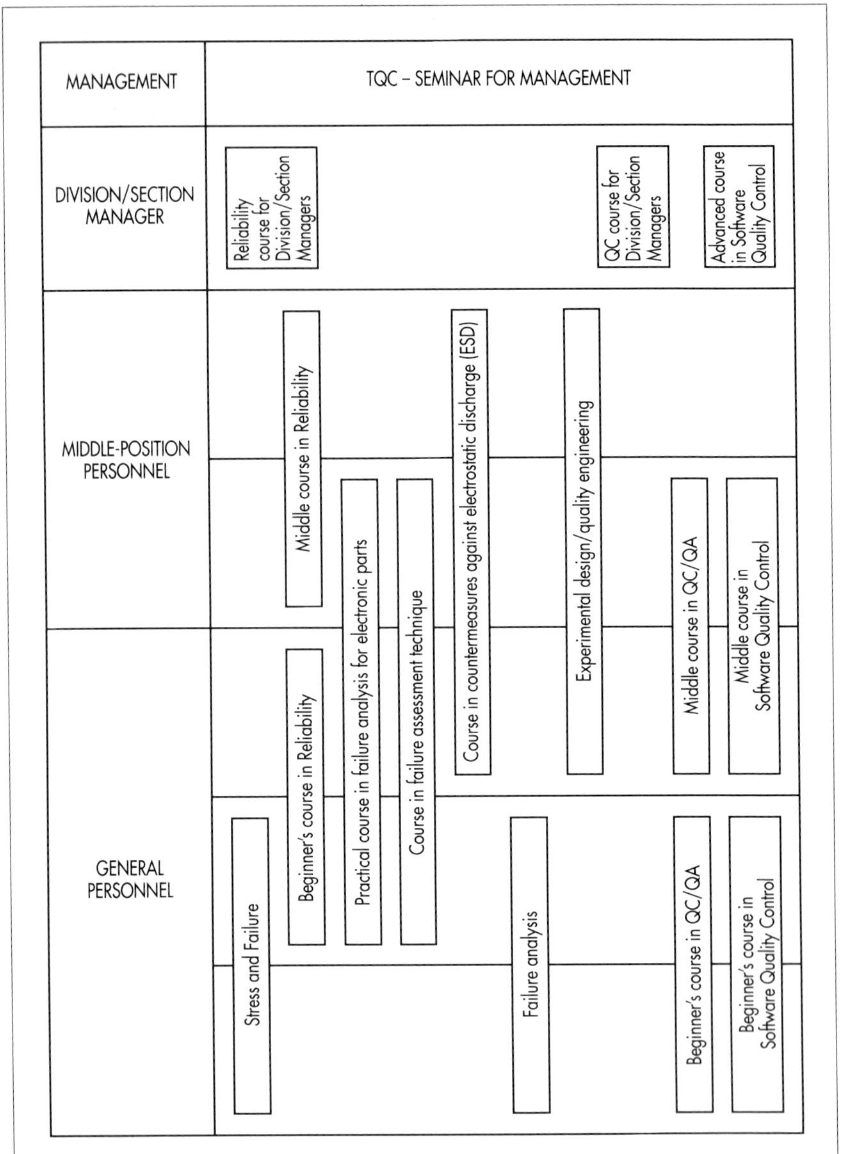


**Figure 2.3** Education system for TQC (Source: Omron Corporation, Kusatu City, Japan)

ment courses (selected employees). Senior managers take a personal interest in the content of the training programme.

Education and training programmes are developed to increase each person's knowledge and skills. The majority of companies encourage their employees to suggest the professional training they wish to undertake. The Japanese firmly believe that better education makes for an improved worker. In one company's training programme, two machine operators were undertaking CAD/CAM training and the plant director had enrolled for a course on gas welding techniques.

The following is an outline of the training given at supervisory level by a mechanical engineering organisation:

- Self development.
- Effective use of time.
- Education of subordinates.
- Labour and personnel management.
- Safety and health management.
- Enhancement of production efficiency.
- Understanding costs.
- Quality consciousness/tools and techniques.
- Process control.
- Maintenance.
- Environmental control.

Both the breadth and depth of this company's supervisor training programme are worthy of comment. For example, in the section on 'Understanding Costs', the supervisors are given detailed information on both fixed and variable costs in their area, plus a general understanding of everything in a production operation that influences costs, e.g., assembly line balancing. This depth of knowledge would be unusual at much more senior levels in European organisations. The breadth of training is illustrated by the information given on what would generally be regarded as other people's jobs. In this way the supervisors are given a total picture of business operations. The majority of the training material was in the form of check-lists that the supervisor was expected to go through before and after taking action.

In general, European companies give little real training at supervisor level. By contrast, Japanese companies regard supervisors as the first level of management and give them the skills to match this responsibility. All senior managers have been through the same training and so have shared knowledge and beliefs. This approach has tremendous power.

Everyone in a Japanese organisation is trained to use the seven original quality control tools,[3] e.g., cause and effect diagram, histogram, check sheet, Pareto diagram, control chart, scatter diagram and graphs, and there was ample evidence (as displayed on the quality notice boards discussed earlier) that these tools were well developed during all stages of production. A variety of training is given in techniques, and presentation and leadership skills to deepen TQC consciousness amongst employees and to improve their problem solving ability. The most intensive training is given to design and production engineers – the samurai of Japanese corporate advance.

In Europe, people are usually trained by one company, often then moving on to another company which benefits from the training and development already done. Companies in the West must be willing to adopt a broad-minded approach and expect to gain as much as they lose from this turnover of people and skills, otherwise there is a danger that investment in education and training will be inhibited. This also frequently results in a failure to develop training strategies; rarely do companies set improvement objectives for training programmes.

In Japan, the quality training programme is typically carried out by staff from corporate and divisional quality departments, engineers, invited lecturers from outside the company and institutions such as JUSE. Almost without exception, the instructors carry out an audit to assess the effective use of newly acquired skills by the people they have trained.

Because of the lifetime employment practises of large Japanese organisations, it is possible for them to invest heavily in a long-term programme of education and training and retraining to develop their employees' capabilities. A programme of initial training prepares them for work with the organisation in terms of mission, philosophy, systems, procedures and job skills. The fact that all new employees, from junior to senior high school and university, begin employment in an organisation at the same time each year helps the organisation of the training and forms the basis for relationships which will last for a very long time.

## Total Productive Maintenance

The concept of Total Productive Maintenance (TPM) is now very much in evidence within Japanese companies. Nippondenso (a subsidiary of the Toyota Motor Group) has achieved a success with TPM, one which every company in Japan is now trying to emulate.

Japanese companies are very much motivated by national prizes. They have the Deming Award which recognises outstanding achievement in quality strategy, management and execution; the Japan Quality Control Medal for companies which have won the Deming Application prize and can thereafter demonstrate five or more years of continuous improvement; the Ishikawa prize for new methods or systems for the modernisation of management and achievement by their application; and the Japan Institute of Plant Maintenance Award for TPM. These awards are perceived as being extremely prestigious. More than one company commented that the TPM award was the hardest of all the awards in Japan to achieve.

There was a range of different views amongst the companies visited on the meaning of TPM. The widest definition was Total Productive Maintenance which encompassed the more narrow term of Total Prevention Maintenance. Therefore, TPM is seen as a total method of management.[4,5] It was Nakajima who developed TPM by combining the key features of preventative, productive and predictive maintenance with TQC, QC circles and employee involvement.

It is clear that TQC and TPM are similar concepts, with the common goal of improving product quality. TPM is considered as an additional driver which is complimentary to TQC. At one company it was suggested that TQC is about *how to* and TPM is concerned with *why*. However, whilst most Japanese companies started TQC some 25 to 30 years ago, the majority of them have only become involved with TPM during the last three or so years. The consensus view amongst companies is that, in their experience, TQC has only a limited influence on machine performance, thus they have introduced TPM to focus on the machines. The condition of the equipment has a considerable influence on the quality of production output and is a key element in the manufacturing of a quality product. The machine needs the input of people to keep it clean and to improve its efficiency and operation. This is the purpose of TPM.

TPM is a scientific, company-wide approach in which every employee is concerned about the maintenance and the quality and efficiency of their equipment. The objective is to reduce the whole life cost of machinery and equipment through more efficient maintenance management. Teamwork is a key element of TPM. It focuses on reducing manufacturing losses and costs, and establishes a system of preventative maintenance over a machine's working life. The emphasis of TPM is to improve the skills of operators in relation to machine technology, and to train and educate operators to clean, maintain and make adjustments to their machine. The training and education of operators is carried out by maintenance and engineering staff. In this way machinery is kept at optimal operating efficiency. The Japanese 5Ss – *seri, seiton, seiso, seiketsu* and *shitsuke* (orderliness, organisation, hygiene, cleanliness and discipline) – are essential activities in TPM. These 5Ss also promote visible management. In addition to QC circles, the Japanese companies operated TPM Circles which tend to focus on the production facility.

In a battery manufacturing company, the three-year TPM programme has seven key steps:

*Step 1* – initial cleaning.
*Step 2* – counter-measures at the source of problems.
*Step 3* – cleaning and lubrication standards.
*Step 4* – general inspection.
*Step 5* – autonomous inspection.
*Step 6* – orderliness and tidiness.
*Step 7* – full autonomous maintenance.

After each TPM step has been achieved, a TPM sticker is issued and placed on a machine. They started TPM to improve efficiency, quality, control and the ability of supervisors. Each employee is given a 20 hour education programme on TPM. The operating personnel devote some eight to ten hours of paid overtime each month to TPM activities, and produce TPM study sheets to educate and involve their peers. Twice a year the best ideas are selected for an award. Silver stars are put on machines which have operated at 100% efficiency during a period of two shift operation. Examples of TPM efficiency improvements made by the company are:

- A 20% increase in machine performance for the battery plate making line.
- A 100% increase on the battery assembly line.

## Just-in-Time

Just-in-Time (JIT) is typically considered by Japanese companies as a key feature of TQC, and so it is worth mentioning how JIT is being developed. Most companies said that their systems were still being developed, and that each day they try to make progress to one-by-one production. The aim in all cases is to eliminate the seven mudas (wastefulness):

- Excess production.
- Waiting.
- Conveyance.
- Motion.
- The process itself.
- Inventory.
- Defects.

This is to ensure that all actions carried out are adding value to the product. The typical cycle employed to eliminate waste is: identify waste, find the cause of waste, make improvements, standardise the improvement to hold the gains (Figure 2.4).

The main factors currently surrounding the deployment of JIT by Japanese companies are:

*Customer service*
The main purpose of JIT is not inventory reduction, but improved delivery of products to customers. The first aim is to reduce waste, and the second is to reduce lead-times.

*Selective use*
JIT is not applied universally to all products and components. The Japanese investigate each piece produced and determine if JIT can be used. For example, the view was formed that the Japanese were finding that JIT was difficult to apply in the case of semi-conductor manufacture. To cater for fluctuating demand, they keep stocks of diffused wafers which are then customised.

*TPM essential*
TPM is crucial to the effective operation of JIT.

*Pilot projects*
When JIT is initially applied, a production/assembly line is chosen as a pilot, where it is developed and refined before being extended to other production/assembly areas.

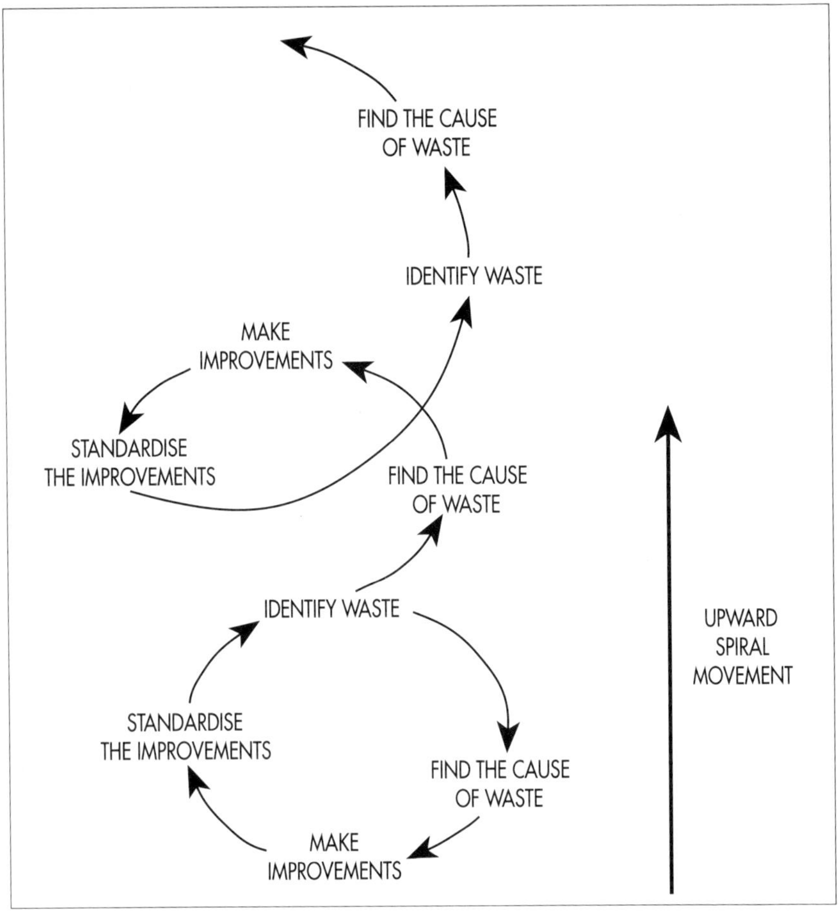


**Figure 2.4** JIT Cycle

*Reduced set-up times*

On the lead oxide grid pasting line of a battery manufacturer, it previously took two operators 30 minutes to change the dies. This has now been reduced to one minute for one person.

*Selective use of Kanbans*

A variety of Kanbans are used – cards, small boxes, Kanban squares and lights. Kanbans are used in only certain parts of the production system, but are also used to order some parts from suppliers.

*Flexible supply chains*

A feature of the production system is that if operator A is being supplied with product by operator B, and his line is not meeting the production schedule, then B stops supplying the product and comes to A's assistance. Some operators are dedicated to the supply of parts to manufacturing and assembly line personnel, and others are responsible for coming to the production line and assisting with emergencies as and when they occur.

*Logistics*

At one company, most suppliers are located within 50 to 60 kilometres of the factory, some within 20 kilometres and others are at a much greater distance. The delivery frequency is dependant upon the quantity and varies daily, every two days and every three days. Another company said they delivered three times a day to original equipment manufacturers, and that on average their suppliers delivered twice a day to them. One company provides a three month forecast to their suppliers, and each day they picked from suppliers' production lines and/or stores for the parts they required. It is interesting to note that at a wire harness manufacturer, this comment was made: "Like it or not we have to be a JIT supplier to Honda, Nissan and Toyota." This comment is typical of that heard in European automotive component suppliers.

The suppliers tend to work to a specific scheduled delivery time on the day in question, and the supply is made in small lorries which deliver right into the factory. There is often a mix of parts from different suppliers on the same truck. This is also the case in the delivery of the finished product to the customer. There is serious traffic congestion in Japan's principal cities, and this has no doubt been attenuated by the transportation of small amounts of goods at frequent intervals. Koshi[6] claims that 43% of motor vehicle traffic in Japan carries freight, and he proposes an underground network for the distribution of goods.

*Supplier audits*

Some parts are supplied direct to the line without inspection, whilst others are inspected. This decision is related to previous quality performance of the parts and the supplier's quality system. An audit to assess a supplier's quality system, manufacturing processes and parts quality is usually carried out. There are three types of audit:

- New supply audit.
- Periodical audit.
- Emergency audit.

The results of incoming inspection and non-conformances found on the line are evaluated each month and fed back to suppliers. Information between customer and supplier is regularly exchanged using computer-aided quality information systems. Quality improvement meetings are usually held once a month with the aim of improving the quality of the product supplied. It is interesting to note that, in general, quality control charts are only requested from suppliers when non-conformances occur. The considerable attention paid to the use of SPC by Western European companies does not occur in Japan. They only tend to use SPC when proving capability and when they are experiencing problems.

*Dual sourcing*
A mixture of single and dual sourcing is employed by the companies visited, and in both cases the relationships with suppliers are long-term. Among the reasons for dual sourcing was to keep some flexibility in a customer's dealings with its suppliers, competition in terms of QCD and the capacity of suppliers. One manager, when questioned about the sourcing of wire harnesses by major Japanese motor manufacturers said: "Of course Honda, Nissan and Toyota employ dual sourcing." These three motor manufacturers divide their requirements between the company and their competitor depending on capacity, demand and schedule. At regular intervals the company confirms its capacity to these three major customers.

*Manufacturing technology*
The Japanese are employing the same machinery and technology as their European counterparts. However, because of activities such as TPM, reduced setting-up time by use of single minute exchange of dies,[7] and integrated materials handling, their machine efficiency and effectiveness is much higher. Not all their machinery and equipment is new. For example, at one company they had die casting machinery which was over 15 years old. A number of their production system improvements have come from developing and applying small, inexpensive customised equipment for the handling and transfer of parts between processes, from processing itself and through the application of mistake-proofing devices. In general, they employ

modern production layouts without huge visible investment; that is, apart from computer-aided design, manufacturing and simulation systems.

*Production system development*
Examples of some of the means employed by Japanese companies in developing their production systems include:

- Product and cellular layout arranged in a U shape. They tend to change the layout to reflect changes in product mix.
- The use of mixed model production and assembly lines.
- The employment of cycle-time conveyors.
- To facilitate easy movement and production flow, operators stand at workstations.
- Considerable efforts are made to ensure smooth production in relation to volume, variety and capacity. They carry out a number of iterations of the sales and production plan to achieve this. Most organisations give their annual production programme to suppliers.
- Where human operation is more efficient and effective than machines, the Japanese will employ it. On a number of occasions it was said that, on some tasks, people could work faster than machines.
- In assembly situations, jobs are kept together in kit form.

## Summary

This chapter is based on the experiences of four study missions of European executives to study TQC in Japanese companies. At the outset of a mission the majority of its members, in general, believed that the Japanese did everything right whereas in Europe everything was done wrong. This view became steadily diluted as the mission progressed. At the conclusion of a mission the consensus view of its members was generally: we did not see anything which European companies could not emulate, and the difference was not so big that we could not catch up – but this would never be achieved.

In most European companies it is known what has to be done, but often managements have not got the real commitment and energy to see it through. This is not helped by worries relating to the future of manufacturing. Japanese companies have a clear vision of the future, and get on with doing simple and essential things very well. Management is the key difference.

A number of simple facts can be learned from the Japanese experience of TQI which we will refer to throughout the rest of this book.

- TQI depends on a systematic approach which is applied consistently throughout an entire organisation.
- There are no 'quick fixes' for the TQC success of Japanese manufacturing companies. Western executives are always on the lookout for the panacea, but unfortunately there is none; this search for the quick fix is often an irritation to the Japanese. Their success is the result of the application of a combination of procedures, systems, tools, improvement actions and considerable hard work and dedication from all employees.
- Senior and middle managers must believe in TQC as a key business strategy and be prepared to stick with it over the long term.
- There must be a permanent, managed process which examines all products, processes and procedures on a continuous basis and develops in all employees the mindset that there is no ideal state. Self-assessment against the Deming Application Prize criteria is an invaluable means of assessing progress.
- Each person should take personal responsibility for the quality assurance activities within their area of control, and quality assurance must be integrated into every process and function of an organisation.
- Planning for improvement must be thorough.
- Improvement is a slow, incremental process. Companies should not expect quick and major benefits from the application of any single method, system and/or tool or technique. To be effective, the quality control tools must be used together.
- There must be a fanatical commitment to finding out what went wrong and putting corrective action into place.
- The concept of TQC is simple. However defining, introducing and fostering the process is a tremendous task and requires total commitment from all employees.

## References

1. Mizuno, S. (Editor), 1988, *Management for Quality Improvement: the 7 New QC Tools*, Productivity Press, Cambridge, MA, USA.

2. Shingo, S., 1986, *Zero Quality Control: Source Inspection and the Poka-Yoke System*, Productivity Press, Cambridge, MA, USA.
3. Ishikawa, K., 1976, *Guide to Quality Control*, Japanese Productivity Association, Tokyo, Japan.
4. Nakajima, S., 1988, *Introduction to Total Productive Maintenance*, Productivity Press, Cambridge, MA, USA.
5. Nakajima, S., 1988, *TPM Development Program*, Productivity Press, Cambridge, MA, USA.
6. Koshi, M., 14 November 1989, "Tokyo's Traffic Congestion can be Unravelled," *The Japan Times*, 5.
7. Shingo, S., 1985, *A Revolution in Manufacturing: the SMED System*, Productivity Press, Cambridge, MA, USA.

# Chapter 3
# The Motivation for Total Quality Improvement

## Introduction

What motivates an organisation's senior management to make product and service quality their number one priority? Unfortunately there is no easy answer to this question.

Changing the behaviour of an organisation is extremely difficult; organisations by their very nature are not meant to change. They are social structures intended to make behaviour more predictable and efficient by rationalising it in the form of a bureaucracy. The bureaucratic sophistication of modern organisations works against change so that organisations usually develop an in-built resistance to it over the years. Organisational change, therefore, can never be spontaneous; there must be good reasons for it to happen.

This chapter examines what causes and motivates organisations to improve product and service quality. It opens by discussing the nature of organisations and the forces which are likely to trigger change. This is followed by a presentation of the quality improvement case histories of five different manufacturing organisations which highlight these forces at work. The chapter concludes by summarising the key change agents and opportunities.

## Organisations and change

Before examining the reasons for organisational change, it is worth considering the nature of organisations and change.

Duncan[1] defines an organisation as "... a collection of interacting and independent individuals who work towards common goals and whose relationships are determined according to a certain structure." His definition emphasises the two interrelated issues that determine the effectiveness of an organisation: commonality of purpose and structure.

In attempting to describe organisational structure and development, many behavioural scientists have focused on contingency theory. This is based on the assumption that there is no universal structure for organisations, that the structure of the organisation is dependent on many factors. Some commentators on the contingency approach to organisational analysis perceive organisations as complex, *dynamic* systems. Much of the contemporary work in organisational psychology makes use of systems ideas. Ackoff and Emery,[2] for example, define a system as a set of inter-related elements or processes. Such hypotheses rest on a fundamental assumption that organisations are open systems that need careful management to satisfy and balance internal needs and to adapt to environmental circumstances.

To understand the behaviour of an organisation, it is also necessary to understand the context within which that behaviour occurred. All organisations engage in activities as a direct response to their environment. But to the disinterested observer an organisation's actual response to environmental cues does not always seem logical. This should not come as a surprise, because judgmental decisions by managers depend on good information and clear perception, both of which are often absent.

Some researchers put this more strongly. Weick[3] suggests that the members of an organisation do not respond to the external environment as it actually is, but to their own perception. The perceived environment may or may not correspond with 'reality'. Morgan[4] examines the concept of organisations as self-producing systems, suggesting that they indulge in a form of narcissism. He likens the charts and diagrams that decorate the manager's office wall to mirrors reflecting an organisation's understanding of itself, but not of reality.

Systems theory in its strongest and most technical form makes detailed proposals about the properties of systems, and the processes involved in system survival, growth and decay. An increasingly large body of thought has reacted against this trend. Silverman[5] is highly critical of the fact that some systems theorists treat organisations as if they were living organisms capable of an independent life of their own. As Silverman points out when discussing the influence of the environment, organisations do not react to their environment; their members do.

Reaction to environmental influences is conditioned by culture. Cultures provide their members with a sense of continuity and identity – basically, a commonality of purpose.

To quote Hampden-Turner:

> *Culture comes from within people and is put together by them to reward the capacities that they have in common. Culture gives continuity and identity to the group. It balances contrasting contributions, and operates a self-steering system which learns from feedback. It works as a pattern of information and can greatly facilitate the exchange of understanding. The values within a culture are more or less harmonious.*[6]

A complementary definition of culture comes from Laurent:

> *An organisation's culture reflects assumptions about clients, employees, mission, products, activities and assumptions that have worked well in the past and which get translated into norms of behaviour, expectations about what is legitimate, desirable ways of thinking and acting. (These) are the locus of its capacity for evolution and change.*[7]

Change is not a simple reaction to external influences, nor is it about 'mechanical' adjustments to structured systems. Organisational change is about groups of people having to modify previously-held beliefs about their world, sometimes forcing them to make uncomfortable compromises. Change can seemingly demand a sacrifice of security and peace of mind. Hampden-Turner[6] visualises organisations facing change on the horns of a dilemma: the need to adapt the organisation to a changing external environment versus the need to integrate the organisation's members internally; the need for periodic change versus the need to preserve continuity.

Change is no more straightforward than the organisations it affects. Whilst in Chapter 1 we separated the concepts of culture, organisation and processes to illustrate the impact of TQI, all three are in fact inextricably linked together. Change affects the entire fabric of the organisation, both hard (processes and structures) and soft (culture).

Given the difficulties involved, why do organisations change? More precisely, why do organisations adopt the practises and values of TQM which require so much change?

## Trigger forces for quality improvement

In an attempt to determine the forces or agents that precipitate the process of organisational change, we analysed the quality improvement case histories of 65 companies. These case histories were prepared, during 1984, as part of the initial public relations activity of the then UK's National Quality Campaign. From this analysis, combined with a study of the literature, discussions with various managers and practitioners, and empirical observation, we identified five 'trigger forces' which played a significant role in causing companies to initiate a process of quality improvement. These were:

- The chief executive.
- Competition.
- Demanding customers.
- A greenfield venture.
- A restart situation.

It was concluded from this piece of research[8] that demanding customers were the most effective force for change. In addition to providing tangible proof of the need for improvement, they also have the potential for actually bringing about a more permanent change in supplier behaviour and attitudes towards quality improvement, by direct intervention and providing the resources to teach and help the supplier to implement and successfully use quality management tools and techniques, and develop an effective quality system based on prevention activities.

(To provide some quantitative evidence on the perceived influence of the five trigger forces, respondents to four separate questionnaire surveys (suppliers to three major automotive companies,[9] chief executives[10] and respondents to the Second and Third International Conferences on TQM organised by IFS Conferences[11]) were asked to identify the factors that provided the motivation to start a formal process of quality improvement. The overall response from 462 different organisations to this question (shown in Figure 3.1) supports the earlier view that demanding customers are indeed the most effective trigger force, closely followed by a need to reduce costs/improve profitability.)

The five trigger forces are now described.

**The chief executive**

Internationally famous authors on the subject of TQM, such as Crosby,[12] Deming,[13] Feigenbaum,[14] Ishikawa[15] and Juran,[16]

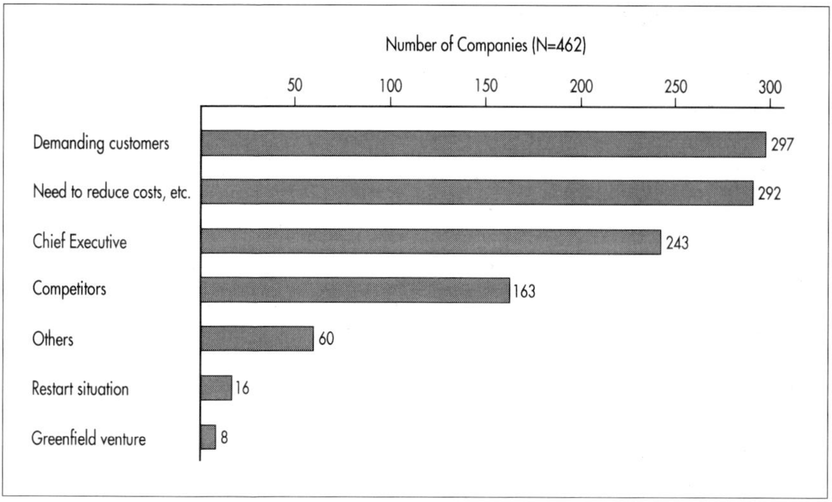


**Figure 3.1** Factors motivating Quality Improvement

are all agreed that unless the chief executive officer takes the lead in a process of quality improvement, any attempts and improvements made by individuals and departments will be transient in nature. In the majority of cases chief executives will not be the sole change agents at work; they are more likely to be one of a combination of change agents (typically in conjunction with competition and/or demanding customers).

The following factors are likely to influence the effectiveness of the chief executive officer as a change agent:

1. Internal factors, such as:

   - Personality and managerial philosophy.
   - Knowledge of TQM in terms of strategy and working practices.
   - Career background.
   - Newness to the job.
   - Issues which impact on organisational culture such as history, ownership, people, location, management style.
   - Attitudes towards TQM within the parent company.

2. Environmental factors which produce a reaction on the part of the chief executive officer and lead to positive action, such as:

- Static profit levels.
- Privatisation.
- Attractiveness (or lack of it) to the market-place in products and services.
- Type of production/operation or service process.
- State of technology – manufacturing, service, information, etc.
- Joint venture agreement, for example, with an organisation more advanced in its strategic quality planning and improvement activities.
- The implications of other strategic actions on which the company has embarked, e.g., diversification of products/services, technological innovation to maintain a market-lead over competitors, and image building to win customer credibility.

**Competition**

Competition is fierce in world-markets and, as outlined in Chapter 1, product and service quality is becoming increasingly recognised as the prime consideration in many purchasing decisions. Fitness for purpose, reliability, durability, consistency and the ability to make continuous improvements are now an essential part of the marketing mix as companies seek ways to effectively differentiate their products from those of competitors and to use quality as the means of increasing their market-share. Most successful companies (in market-share terms) now advertise their product on the basis of quality and reliability rather than price, e.g., the Ford Motor Company's advertising slogan, 'Ford cares about quality', and that of Esso, 'quality at work'. On the other hand, in a number of markets the accepted norm of product quality is a low number of Parts Per Million (PPM) defect level – less than 100 PPM – those organisations failing to perform in these terms do not get orders.

Similarly, new technology, product innovation, attractiveness of products/services, the need for shorter product development times, cost effectiveness, competitive manufacturing and short response times, all combine to form the cutting edge of competition. Such competitive forces compel companies to deploy the latest state-of-the-art management techniques and methods throughout their operations including a process of continuous quality improvement.

There are numerous well-publicised instances where intense competition has been the change agent driving companies to adopt TQM and a process of continuous quality improvement in order to improve product and service quality which satisfies the needs and expectations of the market-place, see the DTI's 'Case for Quality'.[17] The alternative is to go out of business or, at the very least, to lose market-share or withdraw from that market.

The effectiveness of competition as a change agent depends on its severity in a particular market or industry. The tougher the competition, the greater the likelihood that companies will be motivated to improve product and service quality in a bid to maintain or increase market-share. Contributory factors may be classed as either 'defensive' or 'offensive'. These factors should be recognised in determining quality improvement objectives.

Defensive factors result from competitive forces (or changes in them). These include:

- Profits squeezed under competitive pressure.
- Loss of market-share due to entry into the market by overseas competitors committed to TQM and continuous quality improvement.
- Changes in market structure.
- Rationalisation following market contraction due to economic recession.
- Attitudes and behaviour of buyers in export markets.

Offensive factors are linked to some kind of pro-active strategy designed to give a company a competitive lead. These include:

- Breakthrough to improve product and service quality in advance of competitors.
- Product innovation strategies.
- Corporate image-building.

**Demanding customers**

A demanding customer with high product and service quality expectations can lead to a supplier adopting a more formal and effective quality system, developing advanced quality planning methods and systems, and introducing specific quality management techniques to retain the business; especially if the customer provides a major part of the company's business and therefore its profitability.

A number of major purchasers have now articulated and documented what is required from their supplier communities, e.g., the Ford Motor Company's world-wide quality system standard Q-101 and Q1 requirements[18] and the Nissan Motor Manufacturing (UK)'s 'the Nissan way'.[19] These documents describe the fundamentals that must be incorporated into each supplier's quality planning methods and quality system in order to control and improve product quality on a continuous basis. Each supplier is responsible for building on these fundamentals to develop an effective quality system and provide products and services which are free from defects.

Such companies have made it part of their business to help their suppliers improve product and service quality. Their staff spend time in suppliers' plants giving practical advice and assistance. For example, NMMUK has a five-man supplier development team which is working on a continuous basis with its supplier companies. Companies that take an active interest in the business of their suppliers are likely to have a far-reaching effect on the way that suppliers manage quality improvement (see Chapter 6 for details of supply-chain management).

The relevant contributory factors of demanding customers can be divided into the following four groups:

1. Nature of the market, e.g., hi-tech, National Health Service, Ministry of Defence, Civil Aviation Authority, Pacific Rim.
2. Nature of the product/service, e.g., reliability essential, legal/other mandatory requirements, value and/or premium placed on product/service.
3. Influence of major customers, e.g., strength of customer purchasing power, quality audits, and adoption of a quality system standard, such as the BS 5750/ISO 9000 series.
4. Marketing strategy, e.g., upgrading product quality specification to satisfy the target market segment, diversification into markets with more exacting quality requirements, exporting to Japan and her Pacific Rim neighbours.

**Greenfield venture**

A greenfield venture may be defined as:

- A new company.
- A new operational direction for an existing company, e.g., creation of a new strategic business unit as part of a diversification programme.
- An existing company setting up in a new factory.
- An existing company establishing a new operation in its existing premises after a clear-out of plant, product lines, manpower, etc.

A greenfield venture provides the opportunity for the introduction of an effective, unambiguous approach to TQM in a situation where traditions of inadequate conformance levels and performance are not established. In such a situation everything starts from scratch, so there is no vested interest, and no inhibiting behaviour, attitudes and prejudices to overcome. Japanese companies establishing a manufacturing base in the UK have largely followed this approach. It presents an opportunity for management to do all those things the experts and textbooks say should be done to make sure output is right first time, in a cost-effective way from the outset.

The factors influencing the effectiveness of a greenfield venture as a change agent may be divided into two categories: endogenous and exogenous.

Endogenous factors include:

- The personality/background of the chief executive or founding entrepreneur.
- The parent company's quality system and attitude towards TQM and quality improvement, and quality improvement experience and knowledge (if the greenfield venture is the offshoot of another company).
- The reasons for establishing the venture, e.g., marketing strategy, research and development opportunity, organisational restructuring.

Exogenous factors include:

- Geographical location, perhaps in an area that has an attractive infrastructure, e.g., 'Silicon Glen' in Scotland or Telford in Shropshire, England – both being areas in which relatively large numbers of Japanese manufacturers are situated, or an area close to a major customer, e.g., the North-east of England and Nissan Motor Manufacturing (UK) or Derby and Toyota Motor Corporation.
- The nature of the market or industry.
- Customer attitudes and traditional purchasing patterns, e.g., a new supplier sees the adoption of a demonstrable quality system or the use of a particular quality management technique as a way of gaining credibility with potential customers.

### Re-start after hiatus

A hiatus, no matter how brief, brought about by an interruption of the operation of a company or a manufacturing site, may present the kind of opportunity for a new start normally associated with a greenfield venture. Some form of financial or organisational restructuring, including a merger between two or more companies, a take-over by another company and management buy-out, or a major catastrophe, e.g., the fire that destroyed the Manchester Transmission Plant of Eaton Limited in the 1970s, could be the catalyst for such a change agent.

## Trigger forces at work

Each of the following five quality improvement case histories briefly describes the circumstances which precipitated the improvement initiative, identifies the trigger forces at work and summarises their effects.

**Company A (Packaging)**
The company comprises three divisions which, although located on one site, operate as autonomous business units. Company A is a subsidiary of an international group. This case history is concerned with the Open-Top and Plastics Divisions.

*Open-Top Division*
The Open-Top plant was built in the late 1970s, following an approach by a potential customer which, as part of its sourcing strategy, was looking for packaging companies to manufacture two-piece tin-plate cans. At that time this kind of can was a revolutionary advance in packaging technology. The new factory was built in close collaboration with the customer.

There is an almost fanatical obsession with quality improvement within the plant, which in turn reflects the philosophy of this now-major customer. The division has numerous other customers, each with their own specification requirements, but it appears that none are tougher – the general view is 'satisfy them on product and service quality and you can probably satisfy anyone'.

The plant produces around 2.5 million cans every 24 hours – that is a considerable number of cans if something goes wrong. Such magnitude renders attribute sampling of the finished product impractical and to quote the quality control supervisor, "Trying to isolate non-conforming cans would be like trying to find a needle in a haystack." This is why quality assurance resources are concentrated upstream in prevention-based activities, in a bid to get production right first time. The sheer scale of production output, combined with the fact that product reliability is essential – any non-conformities will render the entire batch unusable because the customer cannot afford to run the risk of contamination – instils a collective quality consciousness throughout the plant.

The following story, which is now folklore in the plant, demonstrates the potential effect of customer purchasing power when combined with large-scale production output.

> *A few years ago a competitor delivered to its major customer one batch too many containing non-conforming cans. This customer finally lost patience and promptly cancelled the entire order (58 million cans) and refused to place any further orders with that company. It is no coincidence that most of this competitor's factory is now closed.*

*Plastics Division*

The plastics plant produces plastic injection-moulded products such as containers, tubes, coffee jar caps, aerosol caps, etc. The manufacturing process is by batch production.

In the open-top plant quality improvement is an integral part of the manufacturing process, whereas in this division it is very much based on inspection. This is performed in accordance with an attribute sampling plan based on BS 6001: Part 1[20] which is usually agreed with the customer. The quality system is based on a downstream detection type system and, in addition, an attitude of 'us' and 'them' prevailed – 'us' (production) trying to get output past 'them' (quality control).

The board of the parent company had for some time been aware of the need to improve the plastics division's business performance, and improving product quality was seen as an area requiring action. The company's head of quality services was directed to initiate a formal quality improvement process. An injection-moulded polypropylene snack-food container was chosen as the first quality improvement project. The choice was made because of implicit customer pressure and also because of the results of an analysis which revealed that this product was potentially the division's most profitable, if customer returns could first be reduced and then eliminated. The customer had four suppliers for this product; at that time Company A was the vendor rated fourth and last. As a result the customer placed only small orders, using them as a reserve supplier when other more reliable suppliers could not meet their production requirements in terms of delivery dates and product quantities.

The improvement activity took the form of establishing daily quality records for the product, analysing the data, tackling the problems identified and ensuring the root causes of such problems were eliminated at source. In less than eighteen months, customer rejects fell from 10% of total output to less than 0.3%. The company is now the vendor rated number one by the customer, and the volume of orders has increased to such an extent that extra manufacturing capacity has been established to cope with this. The product is the plastics division's most profitable product and its largest in terms of production volume.

The improvement process for this product has been an undoubted success. However, behaviour and attitudes to quality within the plant have not actually changed, and an intensive quality 'policing' activity is still essential to ensure satisfactory standards of quality conformance. The following is a typical example of this.

> *A whole batch of containers had to be rejected because they were dirty (a critical attribute). The discharge rollers of the moulding machine had not been cleaned, in spite of several requests by the patrol inspector during the course of the shift that this should be carried out. Production supervision did not feel that the rollers required cleaning at that time, even though the output being produced could clearly be seen to be soiled.*

Before summarising the change agents of quality improvement, it is interesting to record the main differences between the Open-Top and Plastics operations:

1. *Production process.* The Plastics Division is engaged in batch production as opposed to the mass production system of the Open-Top plant. There is still a widespread belief within 'plastics' that it is far more difficult to control product quality in batch manufacture because of the inherent variation in the inputs to the process and the process itself.

2. *Degree of integration between production and quality departments.* A prevention-based approach to quality was developed in Open-Top at the same time as all the other management systems, and the quality planning is an integral part of the manufacturing strategy.

Production and quality assurance departments work in tandem, and individual and collective responsibilities are formally documented and rigidly adhered to. There is no such synergy in the Plastics Division, where conformance levels are generally lower, and where quality has still not been integrated into the management and business systems of the organisation. Here quality improvement is being tackled on an individual and project-by-project basis against a background of polarisation between production and quality assurance departments and their personnel.

3. *Age of the organisation.* In our experience, the more established an organisation, the more difficult it is to start and sustain a process of quality improvement. Open-Top was a greenfield venture, whereas the plastics factory had operated without interruption for many years, so that quality improvement was difficult, and required considerable patience, tenacity, and dedication.

4. *Behaviour and attitudes.* To the outside observer there is a clear difference in behaviour and attitudes towards quality improvement on the part of both management and all other grades of employees in the two plants. In Open-Top there was a much stronger belief in the need for quality improvement, with an awareness of the individual's personal responsibility for producing error-free output and pursuing improvement activities.

   In Open-Top, the existence of a demanding customer was the single most important change agent. Three other contributory factors were also involved: the customer's direct involvement from the outset, the fact that reliability of the product was essential, and the sheer scale and cost of potential failure. The plant was a greenfield venture, and this played a major role in the development of positive behaviours and attitudes towards quality. This culture is such that quality improvement is seen as being owned by everyone. For example, a change of operations director (the division's chief executive) had little or no effect on the improvement process, as there was considerable peer pressure on him to conform to the existing organisational culture. In any case, constant pressure from a demanding customer keeps management priorities firmly focused on product quality.

   There was no such pressure in the Plastics Division. The customer in question did not prove to be an effective agent for lasting change, and the quality improvement process undertaken with this product did not spread out to embrace the rest of the product range. In this situation the chief executive was the only potential change agent; and as he did not take the lead, nothing happened.

**Company B (Machine-Tools)**

A machine-tool manufacturer provides an example of a restart situation. This was brought about by the take-over of four British companies by an American conglomerate, and the subsequent consolidation of operations on two manufacturing sites. Whilst the merger did provide an organisational environment which was conducive to change, the actual impetus resulted from the combination of two other change agents – the chief executive, who saw quality improvement as a strategic marketing objective to win orders; and competition, which provided tangible proof of the need for change. The effect of the restart situation was to simplify the process of change and to ease the launch of a formal quality improvement process.

Company B had a strong customer base for its conventional machine-tools, but, in a climate of intense competition and changes in technology, the chief executive saw the need to introduce customers to new computer-controlled models – CNC machine tools. Improving the company's quality performance (traditionally perceived as high) was seen as an integral part of this competitive strategy. A new quality manager (later to become quality director) was appointed to act as the focal point of the improvement process. The quality manager set about establishing a quality system and developing a total approach to the management of quality which would achieve the objective of the marketing strategy.

During this period, one key event also acted as a change agent. Company B negotiated a licensing agreement with a Japanese company to manufacture one of their CNC machine tools. The need to manufacture a new product to the exacting standards of conformance expected by the Japanese necessitated an organisational discipline, and changes in behaviour and attitudes to ensure that the venture was a success. This challenge, which in the early days resulted in widespread panic, helped to focus people's concentration. The licensing agreement was in fact a greenfield venture, which in the quality director's words "...did not carry the sins of the past..."

**Company C (Electronics)**

This company is a subsidiary of a European-based multi-national electronics giant. Structural changes occurred in the company as old products (for which the plant was originally built) and production technologies were phased out whilst new products brought in. These changes formed the basis of a restart situation, out of which resulted a more capital intensive plant with no single dominating technology. This created a number of challenges for management, challenges which extended beyond product quality into areas such as logistics, structural reorganisation, site refurbishment and skills renewal. These new products also faced severe price competition, which led senior management to recognise that measures other than traditional overhead reduction exercises were needed to achieve a better product.

At the time of this rationalisation, old cultural values and attitudes towards product and service quality existed throughout the plant, and quality was perceived as being the quality assurance department's responsibility. There were no measures of performance other than those provided by traditional – and inadequate – administrative reporting procedures. The out-going product quality was controlled by a final inspection process. In short, improvement as a management and business process was not in place.

Two further factors aided the process of change – an influx of new senior managers, and the adoption of quality improvement as a group-wide strategy by the parent company. These new managers brought with them a scientific approach to problem-solving. They found a natural ally in the plant's quality engineering manager, who was a disciple of the Juran philosophy[16] of quality improvement. The senior management team championed a structured framework for improvement activities (as provided by Juran) and became directly involved in leading these activities. Management commitment was reinforced by the parent company's corporate policy statement which instructed all the group's chief executives to pursue company-wide quality improvement as a strategic objective.

**Company D (Computers)**

The company was a manufacturer of computers and peripheral equipment. It provides an example of a business involved in an intensely competitive industry, characterised by high reliability and accuracy requirements, and rapid changes in product innovation, technology and user application.

The relatively short duration of product life-cycles was a major problem and challenge for companies engaged in this type of business activity. Consequently, much of Company D's output consisted of products manufactured in 'one-offs' or small batches. This situation led to a widespread belief that there was little point in attempting to rectify every fault before production – it was better to rely on final test and commissioning procedures in the 'plug and play' mode to achieve fitness for customer use. It was generally accepted that it was not unreasonable for customers to experience 'teething problems' with leading-edge technology.

In the early 1980s, rapid loss of market-share due to intense foreign competition (particularly from the Japanese) coupled with rising customer expectations of 'high-tech' product performance and certain product innovation strategies made Company D realise that it must rebuild its image, change its market strategy, and concentrate on getting new products right first time, from the design stage forward. Customers began to demand higher standards of reliability, and were no longer prepared to wait for computer manufacturers to 'delay' new products; "If Company D cannot deliver there is always someone else who will" became the typical attitude of the marketplace. By way of illustration, at the beginning of the 1980s the usual commissioning period for new hardware was around three months on a customer's site; five years later this was down to one weekend. The result of these changes culminated in the launch of a formal quality improvement process.

Even though the workforce, in general, has been found to have a positive attitude to quality improvement, changing attitudes within the plant has been a long, slow, uphill process. Middle management has probably been the biggest barrier to quality improvement; often they have resisted specific improvement activities, perceiving them to be an erosion of their authority. To succeed, the improvement process has on several occasions had to bypass some of these managers. The lack of continuity at the highest management level has been a major contributory factor to the intransigence of some managers. (The chief executive officer and the strategic objectives of the organisation change frequently at Company D.)

A new product development at one of Company D's other manufacturing locations provides an example of how the process of quality improvement can be facilitated by creating an artificial greenfield situation in an existing production location; this is commonly termed 'brownfield'. A special production unit was established to manufacture a product new to Company D; this, while requiring different production conditions (that is, large-scale process-flow assembly) and new technical management, used part of the existing workforce, who had to acquire new production skills and adopt an improved attitude to quality. This attitude change was helped by the fact that the initial production activities took place behind closed doors in the existing factory, to keep the project secret from competitors, thereby instilling a feeling of elitism among the employees taking part. After product launch, the production unit became 'open', although still separate from the rest of the factory. The internal and external performance measures indicate that the levels of quality conformance have remained high since the launch.

**Company E (Textiles)**

This company is a major textiles group manufacturing fabrics for clothing, home furnishings and industrial use; it is also engaged in the dyeing and printing of fabrics.

In the 1960s, Company E embarked on a series of mergers and take-overs in accordance with the corporate strategy of vertical integration of the time. By the early 1980s, the group, in order to improve profitability and return on capital employed, was faced with the task of having to pull together into a coherent strategic framework, over one hundred separate business units. An outcome of this was the restructuring of the group and the appointment of a new board of directors to manage it – in other words, a restart situation. The company not only faced fierce international competition (especially from the Far East) which depressed its market-share, but also encountered increased competition between high-street retail chains on product style and quality, resulting in customers demanding higher levels of quality conformance. This intense market pressure provided proof of the need for quality improvement. This was reinforced by an internal survey of quality costs which indicated the importance of improved quality performance as a means of increasing profitability.

As the group consists of a wide range of separate businesses operating in diverse markets, there was a danger that market pressure as a change agent could be diluted. In addition, the impact of competition from other manufacturers might vary in accordance with differing market conditions. This kind of organisational environment clearly demonstrated the need for the chief executive to publicly declare his intentions on the issue of quality improvement. The group's chief executive did just that by issuing a quality policy which directed each business unit to formulate a quality management strategy based on Crosby's 'four absolutes'[12] of quality management.

The group operates on what the chief executive calls a 'loose-tight' principle, based on that of Peters and Waterman.[21] This means a small head office providing firm strategic guidelines within which group companies manage their businesses in accordance with their own particular circumstances. The chief executive and his senior head office colleagues are seen as the focus for quality improvement within the group. In addition to the chief executive's on-going public demonstration of his commitment to quality improvement through personal visits, management meetings, in-house videos, etc., a process of top-down awareness was initiated to 'sell' the concepts and requirements of quality management to all the chief executives by means of a series of senior management workshops.

Furthermore, to ensure that quality improvement was not perceived to be an abstract concept, it was announced that all chief executives of group companies would be expected to produce on a quarterly basis a 'Price of Non-conformance Evaluation Report'. This is an example of the use of quality costs as a measure of management performance to assist the process of organisational change.

## Overview of the trigger forces

### The chain of events leading to change

The trigger forces involved in the case studies are shown in tabular form in Table 3.1. In four out of the six cases (for the purpose of this analysis, Company A is counted as two cases), the quality improvement process continues to be developed and advanced in the respective organisations. In the remaining cases the process is stagnant. Clearly the chief executive is an essential change agent if TQI is to be attained, as is evident from the fact that he was an active change agent in the four cases where the quality improvement process is still active. Some form of market pressure, either in the form of competition or demanding customers, was present in all six cases. This reaffirms the importance of market pressure in motivating quality improvement activities in terms of catch-up or breakthrough, and can be summarised as a market-led paradigm of quality improvement.

This suggests the following hypothesis:

> Product quality has been a relatively insignificant issue in post-war British industry. Only during the last decade, and now during the 1990s, has there been any real pressure to improve product

| | Chief Executive | Competition | Demanding Customers | Greenfield venture | Restart | Lasting Improvements |
|---|---|---|---|---|---|---|
| Company A (Open-Top) | √ | | √ | √ | | Yes |
| Company A (Plastics) | | | √ | | | No |
| Company B | √ | √ | | | √ | Yes |
| Company C | √ | √ | | | √ | Yes |
| Company D | | √ | √ | | | No |
| Company E | √ | √ | √ | | √ | Yes |

**Table 3.1** Summary of trigger forces

> and service quality. This is due solely to the belated and increasing realisation in domestic markets that products and services can be supplied with less variation and more consistency than previously thought possible, at no extra cost (to the customer). This is the result of foreign competitors (especially the Japanese) entering British markets with products (which are better designed, more reliable, more consistent, are free from defects, and have more attractive features, etc., than the British-made equivalents). In recent years the performance of foreign competitors has become the standard for domestic industry. This has created a 'quality gap'. The gap is now being closed, but slowly.

This hypothesis can also be applied across the European Community.

Foreign competition acts, therefore, as a catalyst setting-off a chain-reaction: major customers demand a higher standard of product and service quality from all their suppliers. This in turn motivates the chief executives of these suppliers to behave as change agents introducing a process of quality improvement as a strategic (or even, in the short-term, tactical) reaction to changing market conditions.

A fresh start opportunity (that is, a restart situation or greenfield venture) on its own does not appear to be a significant trigger force, but it can make a positive contribution to the process of change by facilitating easier introduction of a quality improvement process and the attendant change in organisational behaviour and attitudes. Two of the cases (Companies B and D) highlight how the creation of artificial greenfield ventures can prove to be a useful tool for facilitating and managing the quality improvement process.

The five trigger forces were originally viewed as separate elements, but it would be more realistic to see them as links in a chain. In the 'market led' paradigm of quality improvement causation competition acts as a catalyst setting off a chain-reaction which enhances quality awareness in the market, resulting in demanding customers and chief executives behaving as change agents, so that awareness and motivation is cascaded down through the organisation as illustrated in Figure 3.2.

### What is a change agent?

The generally accepted definition of a change agent is a person (or group of people) involved in the implementation of change.[22] Our definition of a change agent is a person (or group of people) who causes change as well as implementing it. In both definitions, change

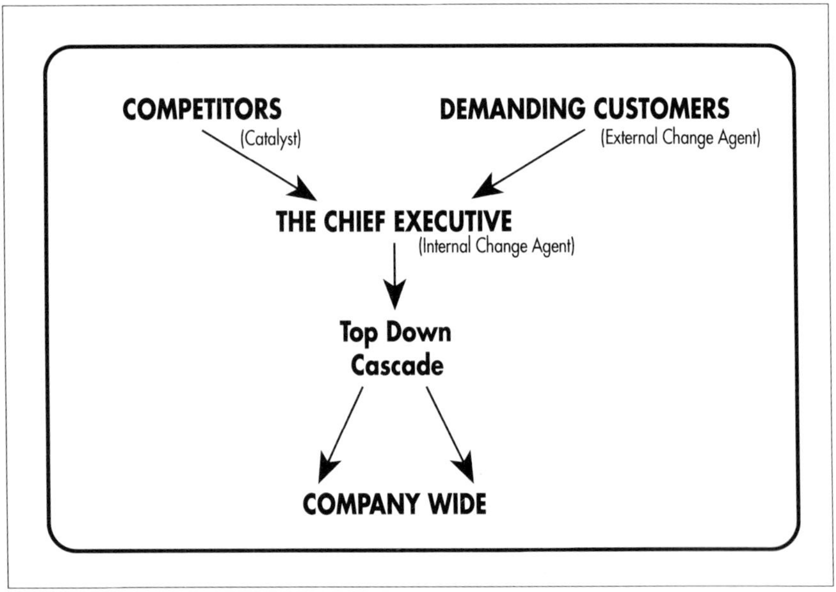


**Figure 3.2** Market-led paradigm of TQI

agents are people. Therefore, the chief executive and customers can be defined as change agents, but competition (that is a dynamic set of market forces), greenfield ventures and restart situations cannot.

If, by definition, competition cannot be regarded as a change agent, it is still the primary catalyst of quality improvement. In addition to the elevation of quality as a component of the marketing mix, a shift has occurred in the fundamental nature of international competition. As discussed in Chapter 1, a more subtle form of competition has emerged that involves the effectiveness of the management of quality improvement activities. For competition to be an effective catalyst, the chief executive must recognise the nature of this paradigm shift and its strategic implications for the business, whether it is the need to fill a potential quality gap in the market-place, or as a means of exploiting a market opportunity. The chief executives of Companies B, C and E, for example, all spotted a potential gap which could be exploited by improving product quality. The Open-Top Division of Company A was established to exploit a specific market niche. However, the effects of competition as a catalyst are probably indirect in many cases. The research evidence indicates that many companies do not bother to monitor the actions of their competitors very closely.

For example, questionnaire survey evidence reveals that, out of 246 automotive suppliers with a formal quality improvement process, only 62% evaluated competitors' products, and only 32% engaged in market research.[9] The structural shift in global competition has resulted in greater pressures on the cost structures of manufacturing industries, and it is this need to reduce costs (an outcome of competitive forces rather than the marketing activities of direct competitors) that is the real catalyst of change in most instances.

Demanding customers can be seen as the key external change agent, and the chief executive as the key internal change agent – and there is little doubt that most companies do react to direct customer pressure. However, demanding customers will not bring about change on their own. The case of Company A (Plastics Division) is an example of where demanding customers failed to effect lasting improvement because the chief executives did not support the improvement process.

Greenfield ventures and restart situations are not in themselves change agents, but opportunities for change acting as positive enhancers of an environment conducive to change. Both can also be used as techniques of change. Artificially creating a greenfield venture, e.g., as in the case of Companies B and D, can be very effective in aiding short-term change, as no major shift in the overall organisational culture is required initially. Any gains will, however, be relatively short lived without company-wide organisational change. Similarly, a restart situation in the form of an organisational or financial restructuring might be used as an expedient method to eliminate traditional barriers to change that typically remain within organisations which have been in existence for a long time, and have well established traditions and culture.

**Categories of quality improvement motivation**

To summarise, the motivators of quality improvement are seen as belonging to one of the following three categories:

1. Catalysts of change
   - Competition.
   - Costs.

2. Change agents
   - Demanding customers.
   - The chief executive.

3. Change opportunities
   - Greenfield ventures.
   - Restart situation.

## Summary

TQI is a form of organisational change. The process is caused by the intervention of change agents perceived as relevant and influential to the organisation. Five forces which may trigger the process have been identified: the chief executive officer, competition, demanding customers, a greenfield venture, and a restart situation. To these a sixth trigger force may be added: changes in the cost structure of the market-place.

The influence of competitors, demanding customers and the chief executive officer are links in the chain of events which leads to change. The interaction between competitors and demanding customers in changing market perceptions of quality has resulted in a structural shift in the market paradigm, which in turn has influenced the perceptions and actions of chief executives.

Greenfield and restart opportunities can contribute to the environment for change, and can be utilised as techniques of change.

The trigger forces can be categorised as catalysts (competition and costs), change agents (demanding customers and chief executives), and change opportunities (greenfield ventures and restart situations).

## References

1. Duncan, W. J., 1981, *Organisational Behaviour*, Houghton Mifflin, Boston.
2. Ackoff, R. L. and Emery, F. E., 1972, *On Purposeful Systems*, Tavistock, London.
3. Weick, K. E., 1977, "Enactment Processes in Organisations," in Staw, B. M. and Salancik, G. R. (Eds.), *New Directions in Organisational Behaviour*, St. Clair Press, New York.
4. Morgan, G., 1987, *Images of Organisation*, Sage Publications, London.
5. Silverman, D., 1970, *The Theory of Organisations*, Heinemann, London.
6. Hampden-Turner, C., 1990, *Corporate Culture: From Vicious to Virtuous Circles*, Hutchinson Business Books, London.
7. Laurent, C., 1990, in Evans, P. (Ed), *Human Resource Management in International Firms*, Macmillan, London.
8. Lascelles, D. M. and Dale, B. G., 1989, "Quality Improvement: What is the Motivation?" *Proceedings of the Institution of Mechanical Engineers*, **203** (B1), 43-50.
9. Lascelles, D. M and Dale, B. G., 1988, "A Study of the Quality Management Methods Employed by U.K. Automotive Suppliers," *Quality and Reliability Engineering International*, **4** (4), 301-309.
10. Lascelles, D. M and Dale, B. G., 1990, "Quality Management: the Chief Executive's Perceptions and Role," *European Management Journal*, **8** (1), 67-75.
11. Dale, B. G., 1991, "Starting on the Road to Success," *The TQM Magazine*, **3** (2), 125-128.
12. Crosby, P. B., 1979, *Quality is Free*, McGraw-Hill, New York.
13. Deming, W. E., 1982, *Quality, Productivity and Competitive Position*, MIT Press, Cambridge, MA, USA.
14. Feigenbaum, A. V., 1983, *Total Quality Control*, McGraw-Hill, New York.
15. Ishikawa, K., 1985, *What is Total Quality Control? The Japanese Way*, (translated by D. J. Lu), Prentice-Hall, Englewood Cliffs, NJ, USA.
16. Juran, J. M., (Editor-in-Chief), 1988, *Quality Control Handbook*, McGraw-Hill, New York.
17. U.K. Department of Trade and Industry, 1989, *The Case for Quality*, Quality, Design and Education Division, Department of Trade and Industry, London.
18. Ford Motor Company, 1990, *World-wide Quality System Standard, Q-101*, Ford Motor Company, Plymouth, MI, USA.
19. Nissan Motor Manufacturing, 1987, Quality Presentation to Suppliers, Nissan Motor Manufacturing (UK) Limited, Sunderland, Tyne and Wear.
20. *Sampling Procedures by Inspection by Attributes, Specification for Sampling Plans*, BS 6001: Part 1: 1991, British Standards Institution, London.
21. Peters, T. J. and Waterman, R. H., 1982, *In Search of Excellence*, Harper and Row, New York.
22. French, W. L. and Bell, C. H., 1978, *Organisational Development*, Prentice-Hall, Englewood Cliffs, NJ, USA.

# Chapter 4
# How Companies Start Quality Improvement

## Introduction

There is no 'sure-fire recipe' for starting and developing the process of TQI. When an organisation's senior management team reach the stage where they want to translate their enhanced quality awareness and organisational need for quality improvement into effective action, they typically seek answers to questions such as:

- What should we do?
- What advice do we need?
- From whom should we be taking advice?
- Can we get unbiased advice?
- Which, if any, of the gurus' teachings should we adopt?
- What TQM packages should we buy?
- What TQM training courses/conferences should we attend?
- Should we call in a management consultant?
- What are the priorities?
- What are the main steps we need to take?
- Should we develop our quality system to meet the requirements of the BS 5750/ISO 9000 series?
- How important is quality system registration?
- What quality management tools and techniques should we use?

Their dilemma is often compounded by their own personal lack of knowledge of TQM and the process of quality improvement, and by a lack of experience in managing organisational change. The overwhelming quantity and variety of available advice, which is often conflicting and sometimes biased or flawed, simply adds to the confusion. It is not surprising then that there is some inertia on the part of senior management teams who are faced with the task of introducing a formal process of continuous quality improvement into their organisations. Smith[1] calls this Total Quality Paralysis. He says: "Just deciding where to begin is so difficult that many never get off the starting block."

In addition, if the bulk of the responsibility for choosing the method and means of getting started rests with the manager to whom the task has been delegated, then there is a certain amount of concern, in terms of future career prospects, about making the correct choice. As a result, there is often a tendency to select the safest option, one which may not be the best for the organisation. The term Total Quality Paralysis is perhaps a little extreme. In our experience, whilst management teams may be uncertain how to structure and start a process of quality improvement, almost all have initiated some kind of quality improvement activity in their respective organisations.

There are four main ways in which an organisation typically starts a process of quality improvement:

1. By applying quality management tools or techniques.
2. By applying the 'received wisdom' of any one of the quality management gurus.
3. By awarding a contract to one of the major quality management consultancies and applying the purchased 'package'.
4. By developing the existing quality system to meet the requirements of the BS 5750/ISO 9000 series, or the quality system standard of a major purchasing organisation.

Organisations are also likely to use a combination of these four methods to get started while they develop for themselves a vision, quality objectives, policy, strategy, an approach and a route map for quality improvement and TQM.

Figure 4.1, based on a questionnaire survey of delegates attending the second and third International Conferences on TQM,[2] shows how organisations got started and/or plan to start on TQM. The most popular method was developing their own tailor-made approach after assimilating the available knowledge on the subject, discussions with other practitioners, visits to other companies, etc. – this is sometimes termed a non-denominational approach – followed by second- and third-party quality system registration and adoption of the teachings of one or more of the quality management gurus.

These four main ways in which an organisation typically starts a process of quality improvement are now discussed.

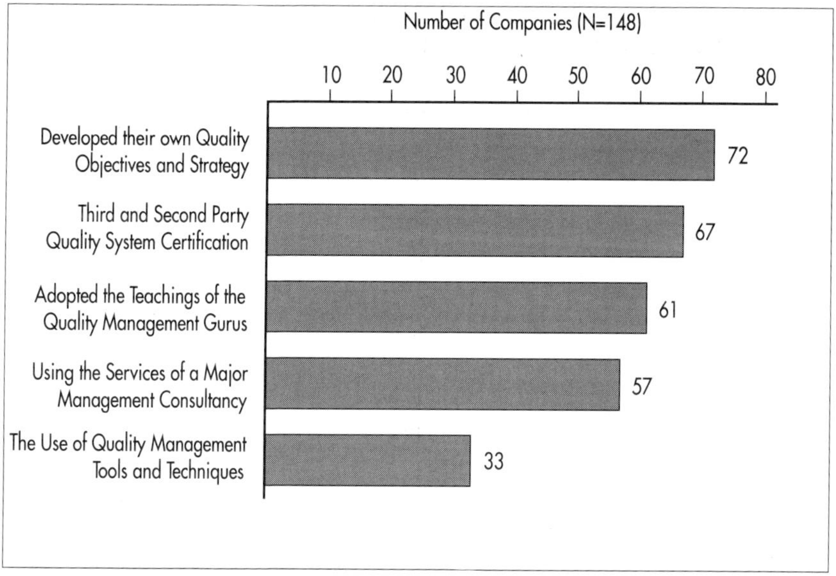


**Figure 4.1** How companies started TQM

## Applying quality management tools and techniques

One approach is for an organisation to adopt and apply a specific quality management tool or technique, and use it to initiate and structure the process of quality improvement. However, irrespective of the TQM approach chosen by the organisation, they will undoubtedly need to use a variety of quality management tools and techniques to assist with the process of quality improvement, from introduction to sustaining its development and advancement.

It is important that organisations do not rush headlong into the use of a plethora of tools and techniques. The techniques used could be those specified as a contractual requirement by a major customer, and may be what management believe the market-place will be expecting in the future. The view may also be taken that the use of a specific technique will give the organisation an edge over the competition. As outlined in Chapter 3, customer pressure can influence the use of specific quality management techniques. For example, among the major automotive companies in the UK, the Ford Motor Company has been the most active in the promotion and use of Statistical Process Control by applying it in its own manufacturing plants, and in requiring that suppliers provide them with statistical evidence of process capability, and demonstrate commitment to the pursuit of continuous improvement. The same is now happening with regard to Failure Mode and Effects Analysis (FMEA) and other advanced quality planning methods.

Dale and Shaw[3] report that when a major customer insists upon the use of a specific quality management technique as a contractual requirement of its suppliers, two phases can be clearly identified in its use. First, the technique is applied by the supplier just to satisfy the demands of the customer in order to maintain the business. During this phase the supplier often resorts to a number of camouflage measures, fakes and ruses to convince the customer that the technique is being applied in an effective and beneficial manner. The emphasis in this phase is on satisfying the customer's paperwork requirements. They go on to make the point that this phase is wasteful in terms of time and resources, but suggest that suppliers appear to need this phase to develop their own awareness and understanding of the technique which is being applied. The second phase begins when the supplier's management team starts to question how they might best use the technique to enhance the company's competitive position. This is when real improvements begin to occur. They also point out that motor industry suppliers appear to have reached this second phase in a shorter period of time with FMEA than they did with SPC, which is indicative of the learning experience.

The approach recommended to suppliers by Dale and Shaw in the use of quality management tools and techniques is that followed by the managing director of a printed circuit board manufacturer: "Forget the Ford Motor Company and General Motors: we need SPC for our own corporate well-being."[4]

## The role of quality management tools and techniques

Quality management tools and techniques fulfil a number of roles. For example:

- Planning for quality.
- Improving the design of the product and process.
- Listening to the voice of the process.
- Improving the process.
- Controlling the process.
- Capturing and documenting quality system data.
- Solving problems.

- Involving people.
- Motivating and promoting quality awareness.

It is important that the management team is fully aware of the main purpose and use of a particular tool or technique. If not, they will soon become frustrated if the technique fails to produce the anticipated result.

We believe that a number of companies use quality management techniques without thinking through the implications of a TQM strategy, or how the concept will be developed and advanced within the organisation. This can give rise to misconceptions and misunderstandings that eventually become barriers to progress. Our evidence indicates that the majority of companies who use specific tools and techniques as the springboard for launching TQM usually single out a specific technique, sometimes at random, and apply it with undue haste and without giving sufficient thought to issues such as:

- How will the technique facilitate the quality improvement process?
- What is the fundamental purpose of the technique?
- Will the technique produce benefits if applied on its own?
- Is the technique right for the company's products, services, processes, people and culture?
- Are we receiving the right advice?
- What organisational changes are necessary to make the most effective use of the technique?
- What is the best method of introducing and then using the technique?
- What are the resources, skills, information, training, etc., required to introduce the technique successfully?
- Has the company the management skills and resources, and the commitment to make it work successfully?
- How will it fit in with, complement, or support other techniques, methods and quality systems already in place, and any that might be introduced in the future?
- What are the limitations, if any, of the technique?

It is important for managers to address such questions when considering the introduction of a quality management technique. Unfortunately, some managers are always looking for techniques as a 'quick fix' solution to the problems facing their organisation at a particular point in time. In general, management teams who are 'technique reactive' tend to be unclear on the concept of TQM and the management of TQI. They often confuse the implementation of a particular technique with TQM and tend to use the technique as an end in itself rather than as a means to an end. This approach is also symptomatic of Western managers' desire for 'quick fix' solutions, as described by authors such as Goldsmith and Clutterbuck,[5] and Peters and Austin.[6]

For example, one company we studied chose as its quality improvement drivers, high-profile 'glory' techniques (namely quality circles, SPC and Taguchi Design of Experiment methods – in that order) as each came into vogue. Because each of these techniques was introduced in an ad hoc fashion, the long-term results were disappointing in spite of the considerable managerial attention, training, resources, energy and money devoted to them. The senior management of this company has now realised that quality improvement cannot be implemented in this way.

**The pros and cons of starting with tools and techniques**

The benefits of starting with a limited number of quality management techniques is that the process of improvement gets under way relatively quickly, employees are involved and feel they are making a worthwhile contribution, quality awareness is enhanced, and behaviour and attitude changes start to happen, albeit slowly. When the process of quality improvement is begun in this way, it is important that:

- The senior management team ensures that they themselves have a good understanding of the techniques to be employed, particularly in the case of statistical methods, so that they can have a common dialogue with people using the techniques.
- Everyone in the company is aware that any tool or technique which is being applied is not an end in itself.
- Improvement objectives and strategies have been developed.
- A quality awareness, education and training programme has been formulated and is in place.
- The organisation has a route map for the techniques which it intends to apply. An example of this, based on the Ford Motor Company, is shown in Figure 4.2.

If the management team is preoccupied with specific techniques, and lacks an adequate understanding of TQM and the quality improvement process, the major danger is that techniques are picked up

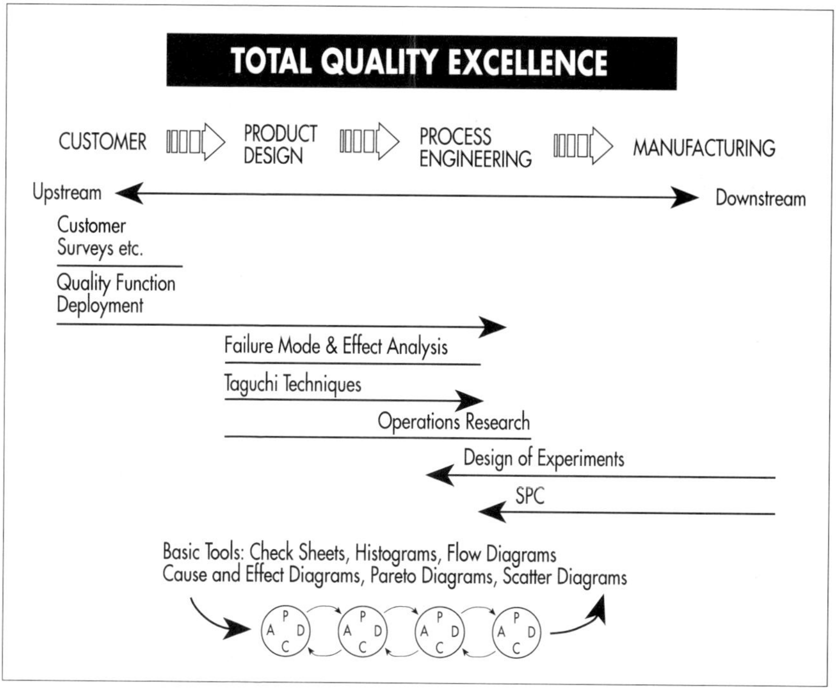


**Figure 4.2** Quality management tools and techniques (Source: Ford Motor Company)

and discarded as fashion changes; an analogy can be drawn from the magician pulling magic balls out of the air or rabbits out of an hat (Figure 4.3). When this happens and the techniques fail to meet expectations, disillusionment sets in and the company will experience considerable difficulty convincing its employees it is serious about product and service improvement.

This, of course, has an adverse effect on the future use of techniques in the organisation. One of the main reasons that companies fall into this trap is that they have unduly high expectations of the benefits arising from the use of a single technique, much of this a result of the hype which often accompanies specific techniques. In general, on its own a single technique will produce only a relatively small incremental improvement. It is only with the cumulative effect of a series of techniques within a strategic framework of TQM that the organisation starts to see long-term benefits from its endeavours (Figure 4.4). Therefore, organisations should never attempt to isolate the benefits arising from any one quality management tool or technique.

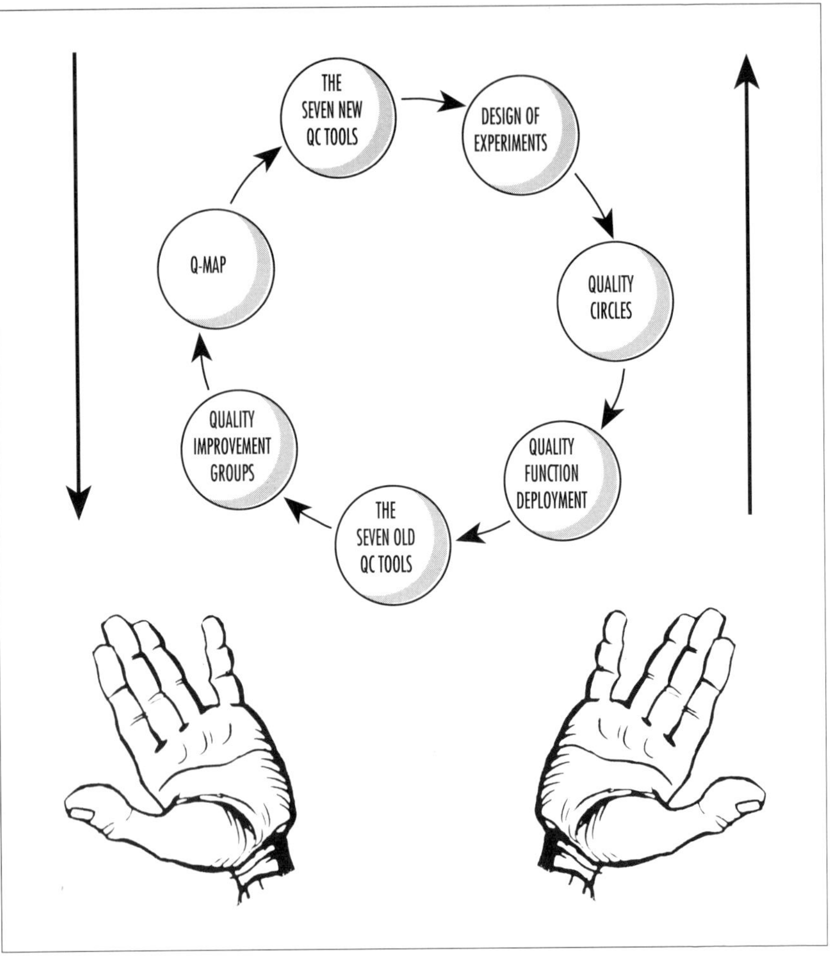


**Figure 4.3** Using quality management tools and techniques

## Selecting which techniques to use

Because of the variety of starting points and motivations for quality improvement, it is not possible to identify a universal implementation plan detailing the order in which specific techniques should be used by an organisation. However, one piece of advice we can offer is that organisations should start with the simpler techniques, such as the seven original quality control tools,[7] and those techniques which are used to facilitate teamwork. Simple tools and techniques can be just as effective as more complex ones. In the West, there is a tendency to

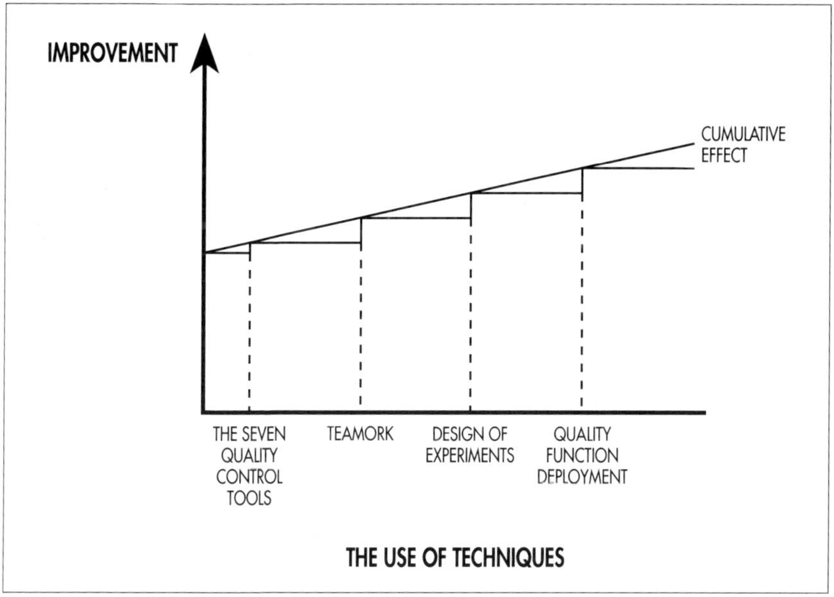


**Figure 4.4** Incremental improvement through tools and techniques

ignore the simple tools and techniques and to use tool and techniques in isolation, a case in point being control charts. Japanese companies, by contrast, tend to use the seven original quality control tools, e.g., cause and effect diagram, histogram, check sheet, Pareto diagram, control chart, scatter diagram and graphs, together, and display the results visibly on a quality control notice board (termed an MQ station). In this way they are not only listening to the process through control charts, but taking action to improve it. This combined use of the seven tools obviously facilitates both problem resolution and improvement action.

When selecting quality management techniques there are two factors which organisations should keep in mind. First, the application of any tool or technique in isolation, without a TQM strategy and long-range management vision, will only provide short-term benefits. For tools and techniques to be effective over the longer-term, major organisational changes in behaviour and attitude are needed. Secondly, no one tool or technique is more important than another – they all have a role to play at some point in the quality improvement process. It is a mistake to single one out for special attention; the Japanese make the point that a warrior should never have a favourite weapon.

Our research evidence suggests that there is an approximate hierarchy of quality management technique usage. Progressing from the basic techniques such as inspection and record-keeping, a company starts to develop quality management systems, and then its approach to continuous quality improvement involves the use of more sophisticated tools and techniques. This corresponds to the levels of inspection, quality control and quality assurance in the evolution of TQM (see Figure 1.6 in Chapter 1). The likely combination of techniques used (what comprises a quality management technique can be open to considerable debate – we have taken a very broad view of technique definition) appears to conform to the following trend:

*Up to 5 techniques*
Inspection of supplies on receipt, inspection during manufacture, inspection prior to despatch, documented quality management policy and quality related records.

*6-10 techniques*
The above techniques plus a selection from: joint quality planning with suppliers, joint quality planning with customers, vendor assessment and rating, quality performance reporting at management meetings, evaluation of customers' purchasing and storage requirements, review of contracts to identify special requirements and the seven original quality control tools.

*11-15 techniques*
Most of the above techniques plus a selection from: product planning, project planning, product testing, evaluation of competitor's products, quality performance reporting to the board, quality costing and quality improvement teams.

*16-20 techniques*
Most of the above techniques plus a selection from: FMEA, mistake-proofing, advanced quality planning methods, reliability improvement programmes, field trials and quality improvement teams.

*Over 20 techniques*
Most of the above techniques plus a selection from: competitive benchmarking, quality function deployment, policy deployment, design review meetings, value analysis, departmental purpose analysis, product life cycle costing and the seven new quality control tools.

Statistical analysis of questionnaire survey data from 300 automotive suppliers[8] indicates that there are significant relationships between technique usage and company size. For example:

- Companies with more than 500 employees are more likely than companies employing fewer than 500 employees to use the following techniques: product planning, quality costing, quality performance reporting at board meetings, FMEA, field trials, quality improvement teams, market research, design briefs and quality circles.
- Companies with more than 500 employees are also more likely to use the techniques of project planning, product testing, evaluation of competitor's product, inspection of product and materials at supplier's premises, reliability improvement programmes, product life-cycle costing, design review meetings, and value analysis.

These findings indicate that larger companies are more likely to use a wider range of quality management tools and techniques than smaller companies. This is not surprising since in small companies there is usually less awareness of quallity, quality skills may be limited, and quality systems are oriented towards a screening activity to catch products with defects before they are shipped to customers. Small firms also have a resource problem and are less likely to have a separate quality assurance department, and managers are generally required to have a wide range of skills. In many cases there is no-one in the company with a working knowledge of quality management tools, techniques or any degree of expertise in quality improvement.

## Applying the 'received wisdom'

Another approach is for an organisation to assimilate and then adopt the writings, guidance and teachings of one of the internationally recognised quality management experts, e.g., Crosby,[9] Deming,[10] Feigenbaum,[11] Juran,[12] and follow whole-heartedly the advice and methods expounded by the chosen expert. The rationale is that this approach provides a coherent framework and gives discipline to the process, enabling people to have a common language and understanding, and consequently helping to reduce confusion. Moreover, because they are all well-proven packages they represent security.

To facilitate this, some companies have purposely opted for the simplest package.

All the companies we have studied that took this approach did utilise the philosophies of the other quality management experts once their quality improvement processes had got under way. This is understandable because none of the experts has all the answers to the problems facing an organisation – despite each guru and their supporters stressing the exclusivity of their approaches/methods. Dale[3] found that Crosby, followed by Juran and then Deming were the most frequently used experts. The approach of Crosby is generally recognised as being the easiest to follow.

As with quality management techniques, whichever expert's package/approach is adopted, it should be treated as a means to assist the quality improvement process and not as an end in itself. It is not unusual for a company to buy a guru's package, and then allow it to fall into disuse once the initial enthusiasm wears off. This is usually because the management team fail to see real-life in-company applications and benefits. For example, in one case study a company had purchased the Juran package,[13] but had not used it to any great extent, and were now introducing SPC. Until it was pointed out, senior management did not see the connection between the use of SPC, and the problem-solving techniques and teamwork skills that could be fostered from the use of the Juran training methods.

The following summaries of the work of Crosby, Deming, Feigenbaum and Juran are based on Dale et al:[14]

**Crosby**

Philip Crosby's audience is primarily top management. He sells his programme to top management, and stresses increasing profitability through quality improvement. He defines quality as conformance to requirements (see Chapter 1).

His programme has 14 steps that focus on how to change the organisation.

1. Management commitment.
2. Quality improvement team.
3. Quality measurement.
4. Cost of quality evaluation.
5. Quality awareness.
6. Corrective action.
7. Establishing an *ad hoc* committee for the zero defects programme.

8. Supervisor training.
9. Zero defects day.
10. Goal setting.
11. Error cause removal.
12. Recognition.
13. Quality councils.
14. Do it over again.

Crosby's approach[9] is based on four absolutes of quality management.

1. Understanding that the definition of quality means conformance to requirements, not elegance.
2. Having a system for quality which is based on prevention.
3. Setting a performance standard of quality which is zero defects.
4. Using a measurement of quality which gets action – the cost of quality.

Crosby does not accept the optimal quality level concept argument, because he believes that higher quality always reduces costs and raises profit. Cost of quality is used as a tool to help achieve that goal. In terms of employee roles, Crosby allocates a moderate amount of responsibility to the quality professional. Top management has the key role. The role of the hourly paid workforce is limited to reporting problems to management. One way that Crosby measures quality achievement is with a matrix – the quality management maturity grid – that charts the stages management goes through from ignorance to enlightenment.[15]

In summary, Crosby is acknowledged as a great motivator of management in helping to initiate the improvement process. His approach is generally regarded as being simple, and easy to follow, and this leads his critics to feel his guidance on how to apply his quality management principles is superficial.[16]

**Deming**

W. Edwards Deming's thesis is that quality improves productivity and competitive position. He defines quality in terms of quality of design, quality of conformance, and quality of the sales and service function. Deming aims to improve quality and productivity, jobs, ensure the long-term survival of the firm, and improve competitive position. He does not accept the trade-off shown in the 'economic cost of quality' models, and says there is no way to calculate the cost of selling defective products to customers, which he believes is the major quality cost.

Deming advocates the measurement of quality by direct statistical measures of manufacturing performance against specification. While all production processes exhibit variation, the goal of quality improvement is to reduce variation. Deming's approach is rooted in statistical methods and he believes that every employee should be trained in statistical quality techniques. A 14-point approach summarises his philosophy for improving quality.[17]

1. Create constancy of purpose towards improvement of product and service, with the aim of becoming competitive, staying in business, and providing jobs.
2. Adopt the new philosophy – we are in a new economic age. Western management must awaken to the challenge, learn its responsibilities and take on leadership for future change.
3. Cease dependence on inspection to achieve quality. Eliminate the need for inspection on a mass basis by building quality into the product in the first place.
4. End the practice of awarding business on the basis of a simple price tag. Instead, minimise total cost. Move towards a single supplier for any one item on a long-term relationship of loyalty and trust.
5. Improve constantly and forever the system of production and service, to improve quality and productivity, and thus constantly decrease costs.
6. Institute training on the job.
7. Institute leadership (see point 12): the aim of supervision should be to help people, machines and gadgets to do a better job. Supervision of management, as well as supervision of production workers, is in need of overhaul.
8. Drive out fear, so that everyone may work effectively for the company.
9. Break down barriers between departments. People in research, design, sales and production must work as a team, to foresee any problems of production or use that may be encountered with the product or service.
10. Eliminate slogans, exhortations and targets for the workforce that ask for zero defects and new levels of productivity. Such exhortations only create adversarial relationships, as the bulk of

the causes of low quality and low productivity belong to the system and thus lie beyond the power of the workforce.

11a. Eliminate work standards (quotas) on the factory floor – instead, substitute leadership.

11b. Eliminate management by objectives, by numbers and by numerical goals – instead, substitute leadership.

12a. Remove barriers that rob the hourly worker of his or her right to pride of workmanship. The responsibility of supervisors must be changed from mere numbers to quality.

12b. Remove barriers that rob people in management and in engineering of their right to pride of workmanship. This means, inter alia, abolishment of the annual or merit rating, and of management by objectives.

13. Institute a vigorous programme of education and self-improvement.

14. Put everybody in the company to work to accomplish the transformation. Transformation is everybody's job.

These points should be used by management to develop its thinking about the way in which business improvement may be made; they should not be regarded merely as a checklist or a set of instructions.

According to Deming, quality management and improvement responsibilities fall to all a firm's employees; top management must adopt the 'new religion' of quality, lead the push for improvement and be involved in all stages of the programme. Hourly workers should be trained and encouraged to prevent defects and improve quality, and should be given challenging, rewarding jobs. Quality professionals should educate other managers in statistical techniques, and concentrate on improving the methods of defect prevention. Finally, statisticians should consult with all areas of the firm.

In summary, Deming expects the managers of firms to change – to develop a partnership with labour, and to manage quality with direct statistical measures without cost-of-quality measures. Deming's approach, particularly his insistence on the need of management to change the organisational culture, is closely aligned with Japanese practice.[15]

It should be noted that a number of Deming Users Groups and Associations have been formed. They are dedicated to facilitating awareness and understanding of his work, and helping companies introduce his ideas.

**Feigenbaum**

Armand V. Feigenbaum was General Electric's world-wide chief of manufacturing operations for a decade until the late 1960s. He is now president of a management consultancy firm, General Systems Co., which designs and installs operational systems in corporations around the world.

Feigenbaum doesn't try so much to create managerial awareness of quality as to help a plant or company design its own system.[16] To Feigenbaum, quality is a way of managing a business organisation. Significant quality improvement can only be achieved in a company through the participation of everyone in the workforce, who must therefore have a good understanding of what management is trying to do. Fire-fighting quality problems has to be replaced with a very clear, customer-oriented quality management process that people can understand and commit themselves to.

Senior management's understanding of the issues surrounding quality improvement and commitment to incorporating quality into their management practice is crucial to the successful installation of Feigenbaum's total quality system. They must abandon short-term motivational programmes that yield no long-lasting improvement. Management must also realise that quality doesn't only mean that customer problems have to be fixed faster; quality leadership is essential to a company's success in the market-place.

Feigenbaum takes a very serious dollars-and-cents approach to the management of quality. He believes that the effective installation and management of the quality improvement process represent the best return-on-investment opportunity for many companies in today's competitive environment.

Feigenbaum's major contribution to the subject of cost of quality has been the recognition that quality costs must be categorised if they are to be managed. In his seminal paper[18] on quality costing he identified three major categories: appraisal costs, prevention costs and failure costs. Total quality cost is the sum of these costs. The goal of Feigenbaum's quality improvement methods is to reduce the total cost of quality from the often quoted 25-30% of annual sales turnover or cost of operations to as low a percentage as possible. Therefore, developing cost of quality data and tracking it on an on-going basis is an integral part of the process.

Feigenbaum says that management must commit itself to:

- Strengthening the quality improvement process itself.

- Making sure that quality improvement becomes a habit.
- Managing quality and cost as complementary objectives.

In summary, though he does not espouse fourteen points or steps like Deming or Crosby, his approach is not dissimilar: it simply boils down to managerial know-how.

**Juran**

Joseph Juran has perhaps made a greater contribution to the quality management literature than any other quality professional. Part of his thesis is that companies must reduce the cost of quality. This is radically different from Deming. Deming ignores the cost of quality while Juran, like Crosby and Feigenbaum, claims that reducing it is a key objective.

Juran defines quality as fitness for use, which he breaks into quality of design, quality of conformance, availability, and field service. The goals of Juran's approach are increased conformance and decreased cost of quality, and goals are set yearly in the objective-setting phase of the process. Basically, his work is in three sections: one attacking sporadic problems, one attacking chronic problems, and an annual quality programme, in which top management participates, to develop or refine policies. Juran defines two major kinds of quality management – breakthrough (encouraging the occurrence of good things), which attacks chronic problems, and control, (preventing the occurrence of bad things), which attacks sporadic problems.

Juran also allocates responsibility among the workforce differently from Deming. He puts the primary responsibility on quality professionals (who served as consultants to top management and workers in Deming's programme). The quality professionals design and develop the approach, and do most of the work. While granting the importance of top management support, Juran places more of the quality leadership responsibility on middle management and quality professionals. The role of the workforce is mainly concerned with specific team-based improvement activities.

In summary, Juran emphasises the cost of quality, because the language of top management is money, and he recommends cost of quality for identifying quality improvement opportunities and developing a quality cost scoreboard to measure quality costs. Juran's approach is more consistent with Western management practices – he takes the existing management culture as a starting point and builds quality improvement systems from that baseline.[15]

**Is there a common theme?**

The ways of approaching quality management suggested by Crosby, Deming, Feigenbaum and Juran are variations on a theme – the essential difference is the focus of their approaches. Several commentators, for example Fine,[15] Main,[16] Geeruliet,[19] and Bendell,[20] have compared and contrasted the approaches of the four gurus. Broadly speaking, the teaching of these four gurus can be characterised by the main focus of their approach, as follows:

*Crosby* – company-wide motivation.
*Deming* – statistical process control.
*Feigenbaum* – systems management.
*Juran* – project management.

Is there any sure-fire TQM route for those seeking the 'philosopher's stone'? Are these really conflicting or mutually exclusive choices? The short answer is that there is no sure-fire solution and that these four approaches are compatible.

Advocates of each guru are apt to claim that their man's approach is the only one likely to work. This is an arrogant and myopic stance to adopt; each approach has its strengths and weaknesses, and they are all proven packages. It is also worth remembering that all four gurus are consultants, and so it is in their business interests to distinguish their approaches from that of their peers and appear to have all the answers. McBryde[21] says that the golden thread running through the philosophies of all four (and other, unnamed) gurus is the concept of adopting quality as a fundamental business strategy permeating the culture of the entire organisation. Fine[15] concludes that the teachings of Crosby, Deming and Juran (Feigenbaum is not included in his comparison) have four points in common:

- The importance of top management support and participation.
- The need for workforce training and education.
- Quality management requires careful planning and a philosophy of company-wide involvement.
- Quality improvement programmes must represent permanent, on-going activities.

We suggest that any person interested in learning about the approaches and methods of any of these quality management gurus goes to the original works rather than to those of their disciples.

Irrespective of approach, package or system, it is the senior management team's commitment to making quality improvement work, and to understanding what they are doing, that is the key to long-term success.

**The Japanese wisdom**

It would be an oversight not to mention the TQM wisdom and experience which is undoubtedly possessed by the Japanese. In recent years, the work and ideas of a number of Japanese quality management experts have been published in English and, therefore, made more accessible to the majority of Western managers and scholars. They include Imai,[22] Ishikawa,[23] Mizuno,[24] Nemoto,[25] Shingo,[26] Suzaki[27] and Taguchi.[28]

The Japanese typically treat the management of quality as waste reduction in its broadest sense, and their goal is continual improvement towards perfection. They allocate responsibility for quality improvement among all employees. The shopfloor workers are primarily responsible for improvement activities. Managers do less on maintaining, and more on planning and improving. At the highest levels, the emphasis is on breakthrough activities and on teamwork throughout the organisation.

The 'loss function' concept of Genichi Taguchi[28] typifies the target-focused approach of the Japanese quality management philosophy. Taguchi defines the quality of a product as: "... the loss imparted to society from the time the product is shipped." Among the losses he includes consumers' dissatisfaction, warranty costs, loss of reputation and, ultimately, loss of market-share.

Taguchi maintains that a product does not start causing losses only when it is out of specification, but when there is any deviation from the target value. Moreover, in most cases the loss to society can be represented by a quadratic function, i.e., the loss increases as the square of the deviation from the target value. This leads to the important conclusion that quality (as defined by Taguchi) is most economically achieved by minimising variation, rather than by strict conformance to specification.

There are a number of now-familiar concepts associated with Japanese TQC. These include:

- The next person or process is the customer.
- Listening to the voice of the customer.
- Commitment to improvement and perfection.
- Elimination of waste.

- Insistence on compliance to procedures.
- Correcting one's own errors.

Various practices facilitate TQC in Japanese corporations (see Chapter 2), including:

- Policy deployment.
- Quality function deployment.
- Use of statistical methods (simple and advanced).
- Daily management control.
- Housekeeping.
- Successive and self-check systems.
- Mistake-proofing.
- Detailed quality assurance procedures.
- Closed-loop defect analysis and corrective action procedures.
- JIT.
- TPM.
- Quality control circles

In the West, there is a fascination with Japanese management styles and techniques. However, many admirers of Japan Incorporated stand accused of wanting to copy directly without properly assimilating and understanding the lessons of Japanese business. Snowdon[29] calls this the 'factory tour' or 'read a book about Japan' approach. As he points out, the Japanese have developed their present industrial structures over a period of four decades – so it is not surprising that it takes more than a few days to assimilate the results. There is also the added danger of essential issues being confused with cultural issues. Dorsky[30] cautions readers not to blindly emulate Japanese practices; Japanese problems and responses may, necessarily, be different from those of the West. A number of writers on the Japanese approach to quality management, e.g., Lee and Ebrahimpour,[31] McMillan,[32] Schonberger[33] and Trevor,[34] agree that there is one fundamental lesson to be learned: it requires commitment, patience and tenacity.

## Applying a consultancy package

Some companies (usually large concerns) decide to adopt the programme of one of the major management consultancies on the grounds that it is a self-contained package which can be suitably customised for application throughout their organisation. Some are very comfortable with consultants, others not so comfortable.

It is important for a company to understand that the use of a consultancy organisation does not relieve the board of their own responsibilities for TQM, it is their responsibility to own the improvement process and they must exercise leadership. The company should never allow the consultant to become the TQM champion and the company guru on TQM. A key part of consultancy is the transfer of skills and knowledge. When the project is complete, the training, guidance, etc., provided by the consultant must remain within the organisation to further the process of improvement. The consultant should be perceived by the organisation as an implementation tool, and not as an initiator of TQM and the improvement process. More often than not the consultant is learning on the job, and any ideas, proposals and decisions should be scrutinised carefully by the TQM steering committee for their applicability to the company's operations.

Quality management consultants bring their expertise to the company and provide the resources, experience, disciplines, objectivity and catalyst for getting the process started. The consultants are usually involved in a wide range of activities from planning through to training, project work, and the implementation of specific improvement initiatives.

There is a myriad of consultancies offering a variety of TQM products and packages, and not all of them will suit every organisation. It is likely that organisations, particularly large ones, will use more than one consultancy as they progress in TQM.

A company intending to use a consultancy organisation must carefully consider the method for its selection and, sometimes, the individual consultant who is most suited to their needs. This generally involves a presentation of their approach to senior management and other interested parties. There are a number of factors to be taken into consideration including:

- The personality of the consultants, and the perceived interaction with the people with whom they will work. The presentation and proposal made.
- Previously published material.
- Availability of educational material and supporting systems and tools.
- The reputation and track record of the consultancy organisation and individual consultant with existing clients.

- Their knowledge of TQM and its practical application in similar or related companies – not just in consulting and/or teaching, rapport and ability to communicate with staff at all levels in the organisation.
- Grasp of the client organisation's culture and management style.
- The extent to which the consultancy is prepared to assist in tailoring the package to suit individual client needs.
- Commitment to support the client organisation.

The decision to use a consultancy organisation is usually made by the chief executive with support from the quality director. It is unwise for the consultant to assume that other board members will contribute more than vocal support to the process of improvement.

The company needs to carefully consider what they are buying from a consultancy organisation. It is often difficult to define in precise detail what is required in a TQM assignment, with the consequence that the terms of reference are vague. This sometimes results in a difference between what was ordered and what is delivered; wrangling over the deliverables from a TQM consultancy contract is a major detractor in a process of continuous improvement. The company should also be aware that a TQM assignment is not used to open the door for other consultancy work in problem areas such as manufacturing management, human resources, organisation development and accountancy. The easiest way of selling consultancy is on the back of a short-term successful assignment; this might have a negative influence on the long-term success of TQM.

A major complaint about consultancies is the use of 'off-the-shelf' packages that fail to maximise client involvement, do not reflect the client's business process, business constraints, and the use of prescriptive words and terms which do not suit the culture of the client organisation.

For those considering the utilisation of an external consultant, three issues need to be addressed and agreed on:

1. Clear terms of reference, specifying the expected benefits to the organisation from the consultancy project, with tangible objectives, milestones and time-scales.
2. The precise nature of the relationship between the client management team and the consultant. The management team will need to consider the precise form of consultancy input required, and identify success criteria for the project.

3. The mechanism for implementing the improvement strategy and managing the change process, together with the resources required. The issues involved include the role of the senior management team, the amount of time and energy individual senior managers are able or prepared to commit, and who might assume the day-to-day role of project co-ordinator.

## Developing a quality system

A quality system should define and cover all facets of an organisation's operation, including identifying and meeting the needs and requirements of customers, design, planning, purchasing, manufacturing, packaging, storage, delivery and service, together with all relevant activities carried out within these functions. It deals with organisation, responsibilities, procedures and processes. Put simply, a quality system is good management practice. If it is to be effective, the quality system must be developed against a reference base on which its adequacy can be judged and improvements made. This reference base is a 'quality system standard'.

### Origins of quality system standards

The early standards were provided by major purchasers to their suppliers. These standards were customer specific and designed to be used only in contractual situations in those industries for which they were designed. Each purchaser developed their own methods of assessment, which involved visiting the supplier to examine the degree to which their operating procedures and systems followed the requirements of the standard – known as *second party certification*.

Most of the current quality system standards evolved from military standards. A UK standard for quality systems was first published in 1973 by the Ministry of Defence (MOD) – the DEF-STAN 05 series (05-21 to 05-29).[35] These were virtual copies of the American-derived Allied Quality Assurance Publications (AQAP-1 to AQAP-9),[36] used by NATO in defence procurements. The MOD used the DEF-STAN 05 series to approve potential suppliers and to audit current suppliers in contractual situations.

It was a requirement that a supplier developed its quality system to meet the clauses set out in these standards as a prerequisite for inclusion on the list of MOD assessed contractors. In 1985 the 05 series was replaced by the AQAP standards. As from September 1991, the MOD has mainly relied on third-party assessment against the ISO 9000 series of standards. They only assess suppliers and contractors who are outside the current scope of the accredited certification bodies. This type of situation relates to specific military applications such as aircraft construction, ammunition and explosives, packaging and computer software. A new set of defence standards (the 05-90 series, 05-91 to 05-95), which includes the ISO 9000 series plus special military purchase requirements, will be used to audit suppliers in contractual situations.

During the mid-1970s there was a proliferation of quality system standards produced by a variety of second- and third-party organisations. The Warner report[37] stressed the need for a national standard for quality management systems to reduce the number of assessments to which suppliers were being subjected by their customers. The report emphasised the shortcomings and fragmented nature of the British system of standards. It recommended that British Standards be produced to provide the single base document for quality systems. Subsequently in 1979, the BSI issued BS 5750 *Quality Systems*.[38]

In 1987, the series of international standards on quality systems was published by the International Organisation for Standardisation (ISO). These standards, whilst reflecting various national approaches and international requirements, are based largely on the 1979 version of BS 5750, and the eight or so years of UK user experience. The text of these international standards has been approved as suitable for publication as a British Standard without deviation – BS 5750: Parts 0 to 3. The ISO 9000 series have now been adopted by CEN (the European Committee for Electrotechnical Standardisation) and CENELEC (the European Committee for Electrotechnical Standardisation) as the EN 29000 series, thus harmonising the European Community approach to quality systems. In 1991 the series was extended to cover software and services.

In the UK during the last decade or so, many organisations have developed their quality systems to meet the requirements of the BS 5750/ISO 9000 series or of an equivalent third-party assessment scheme. Many more have developed an approach to quality management based on the quality system requirements of a major purchasing organisation, e.g., Ford Motor Company's Q-101 Quality System Standard.[39]

## The ISO 9000 quality system series

In simple terms, the objective of the ISO 9000 series is to give purchasers an assurance that the quality of the products and/or services provided by suppliers meets their requirements.

This series of standards sets out a definitive list of features and characteristics which, it is considered, should be present in an organisation's management control system through documented policies, manuals and procedures. The aim is systematic quality assurance and control. It is the broad principles of control, in general terms, which are defined in the standards, and not the specific methods by which control can be achieved. This allows the standard to be interpreted and applied in a wide range of situations and environments, and allows each organisation to develop its own system within the guidelines of the standard.

The ISO 9000 series was developed by the engineering sector of industry, and consequently many of the terms and definitions are engineering-based. They therefore require interpretation for the specific needs of individual organisations, and this has led to problems in some sectors of manufacturing industry, and in the majority of non-manufacturing organisations.

Despite these types of difficulties, it has been applied in a wide variety of manufacturing situations, and registration is being achieved in an increasing number of non-manufacturing environments – banking, legal, consumer, transport, hotel and catering, recruitment, marketing services, and education.

Registration to ISO 9001, ISO 9002 or ISO 9003 is a useful foundation leading to the development of a quality system to meet the independent system requirements of customers.

A number of major purchasers use this registration as the 'first-pass' over a supplier's quality system. They will take the ISO 9000 series as the base, and only assess those aspects of the system which they believe are important.

A documented quality system provides an effective managerial framework on which to develop a company-wide approach to continuous quality improvement. Possession of such system certification may demonstrate an organisation's capability to deliver to the customer conforming products and services, but it does not guarantee the management team's intention to do so. Indeed, possession of certification often results in a sense of complacency, in particular, after successful assessment of the system.

## The limitations of the ISO 9000 series

1. *Many organisations and managers have inflated views of the ISO 9000 series.* Often it is a result of the hype generated by those selling advisory services. Such views can lead to high expectations

of what the standard can achieve, which in the long-term may do it a dis-service. The following comments are typical:

– "BSI quality recognition is prized nation-wide because it is known to be difficult to achieve the high standards required by its impartial testing procedures."
– "... it will give the car-buying public a guarantee of complete satisfaction or their money back. What it aims to achieve is the world's coveted benchmark of quality: BS 5750. ... it is a standard that is recognised as being truly superb and is a move that no other rival car maker can afford to ignore."

Some people are confused about the relationship between the ISO 9000 series and TQM. They are not alternatives – a quality system is an essential feature of TQM. However, some organisations see ISO 9000 series registration as the pinnacle of their TQM achievement. It is our view that on its own the ISO 9000 series has a limited impact on the total improvement operation of an organisation simply because it does not get at the root cause of problems. Most problems are resolved at branch level, and this is a failure, at least in the UK.

2. *In general, the ISO 9000 series tends to measure the effectiveness of documentation, paper work and procedures.* Whilst recognising that the requirements of the ISO 9000 series are aimed primarily at preventing non-conformity at all stages from design through to servicing, registration does not as a matter of course imply that such non-conformities will not occur. The standard is not prescriptive as to the means of prevention. Detection methods which rely heavily on inspection techniques, human or mechanical, would appear to satisfy the standard in many aspects. The standard does state that corrective action procedures should be established, documented and maintained to prevent recurrence of non-conforming product, and that the system is maintained and developed through the internal audit and management review, but we remain unconvinced that improvement is an explicit criterion by which on-going registration is monitored, or indeed a basis for reassessment and registration.

3. *ISO 9000 system series registration, or for that matter any other quality system certification or approvals, will not prevent a supplier from producing and delivering non-conforming products to its customers.* There is a clear difference between registration and capability, which organisations need to recognise. Product and/or service

quality is determined by the individual organisation and not by quality system standards.

Proof of this difference is provided from a recent investigation of the initial samples submitted by some 300 suppliers to the Leyland Daf assembly plant carried out by Mone et al.[40] It is a contractual requirement that all suppliers are registered to the ISO 9000 series or its equivalent. Over 30% of initial samples were found to be non-conforming.

Suppliers have a habit of doing what their customers want, and there is no doubt that many organisations have achieved ISO 9000 system series registration to provide documented proof that they have an adequate quality system in place just to satisfy the demands of their major customers. Any resulting gains made will be short-lived if registration is perceived as a contractual condition rather than a foundation for on-going improvement. It is not unusual for suppliers to adopt the practice of 'stratified quality assurance' by grading their product or services according to their perception of customer purchasing power or quality system requirements. Such organisations miss the point that a documented quality system can be developed into an effective company-wide capability to satisfy all their customers' requirements.

4. *There is no evidence that companies registered to the ISO 9000 system series exhibit a better quality performance, and have a more positive approach to continuous improvement, than those which are not registered.* Registration does not differentiate competing organisations. Every competitor will conceivably have the same entry qualification; the winners will be those who have a dedicated commitment to never-ending improvement. An organisation does not achieve superior performing company status merely by ISO 9000 system series registration.

   Registration to the ISO 9000 system series should be viewed as the minimum requirement. The objective should be to develop and improve the quality system.

## A foundation for improvement

1. *A quality system is a fundamental pillar in an organisation's approach to TQM.* The guidance provided in the 20 clauses of ISO 9001, and the independent assessment surveillance, is an indisputable aid in developing and maintaining the procedures, controls and discipline required in a system.

2. *There is little doubt that the preparation of systems, procedures, working instruction, etc., to meet the requirements of ISO 9000, will have a beneficial effect on a company's performance in terms of improved process yields, reduced levels of non-conformance, etc.* However, the underlying mechanisms of the ISO 9000 system series are such that they will tend to reinforce the status quo. The ISO 9000 system series is designed to produce consistency both in actions and products. It is not unusual to find companies with a consistent performance and a high level of non-conformance.

   Only if there is strong leadership and a real commitment to improvement will the improvement cycle be triggered. Some organisations have done this by widening their quality systems management review meetings (which deal with issues such as quality audit and corrective action; production rejections, concessions and corrective actions; waste levels; supplier performance/concessions; customer complaints; and market trends and requirements) into quality improvement steering-group meetings.

3. *The ISO 9000 system series should be considered as the minimum requirement, and a first step on the road to quality improvement.* Organisations should aim to have a quality system which surpasses the standard's requirements. A quality system which meets the requirements of the ISO 9000 series should in no sense be regarded by senior management as the pinnacle of achievement. All it says to the outside world is that the organisation has controls, procedures and disciplines in place. Without a documented system there are no reliable data to monitor the process of quality improvement. The organisation should treat it as a precursor to developing its approach to TQM.

## Developing a tailor-made route map

A variation on the four approaches and their various combinations is to absorb the 'received wisdom' and the experiences of other companies, and extract the ideas, methods and tactics which are appropriate to the particular circumstances, business situation and environmental culture of the organisation. In our experience, organisations starting with any of the four approaches will eventually use this method. A feature of organisations following this approach is that senior management will have visited other companies with a reputation for being 'centres of excellence' to see at first hand the lessons learned from TQM and become involved in meetings relating to TQM with executives of like mind from different companies. They will also have attended conferences frequently, and be generally well-read on the subject.

We believe that when getting started on the quality improvement process it is always beneficial for organisations to establish contacts with others that have a reputation for excellence in systems and products. There is much to be said for learning by association. In our experience, companies working with or competing directly against companies with advanced management processes develop their knowledge of TQM and quality improvement at a fast rate.

In our view this tailor-made route, or what is termed 'non-denominational' in the key excellence indicators of the leadership criteria of the Malcolm Baldrige National Quality Award, is the one which organisations should follow.

## Some examples of getting started

We have studied several examples of how companies developed a strategy of quality improvement and achieved positive results. Three different approaches are now discussed.

**Company 1**

This company's quality improvement strategy had three main objectives, covering distinct areas of activity:

1. To improve management skills, and particularly quality skills, by an intensive training programme and exposure to the appropriate quality and problem-solving techniques, using both internal and external sources of expertise.
2. To improve the management of control activities by introducing an internationally recognised quality systems standard and a plant-wide quality audit.
3. To develop a programme of increased employee quality awareness and involvement through a publicity campaign and education programme.

These three areas were considered to be an integral part of a total approach to quality management, and as such had to be tackled simultaneously. The three areas were termed: breakthrough, control, and awareness.

- *Breakthrough* focused on more effective product management of technological improvement by improving the company's methods of project identification, organisation and reporting; resource management; and quality training. Juran's structured training package[13] was used to facilitate this, and at the time of the study over 250 people had received training in this way.

- *Control* focused on more effective management control of processes and services by improving systems, communications, formal instructions, and procedures. The key was seen to be internal quality audits and the Allied Quality Assurance Publications (AQAP) series of Quality System Standards.[36]

- *Awareness* focused on more effective actions to promote quality awareness and stimulate the involvement and motivation of all employees through a plant quality steering group, quality awareness sessions, quality circles, quality and performance indicators, and quality posters. The key was seen to be involvement.

The lesson to be learned from this company's approach is the way in which the activities and objectives of the quality improvement process were divided into concise, clearly defined and manageable sections.

**Company 2**
At this company, the quality improvement strategy followed three sequential phases: awareness, measurement, and process management.

- *Phase one was awareness.* The first thing the company did was to establish a formal world-wide quality improvement management structure. For example, each of the operating divisions, e.g., sales, customer engineering, administration, in addition to the manufacturing plants, appointed a quality improvement co-ordinator. A top-down quality awareness education programme, similar to the cascade models advocated by Crosby[9] and Smith,[1] was carried out. All the directors received quality awareness education first, and then during the next nine months so did every manager in the company. Managers were taught to use a quality improvement methodology and encouraged to apply it to projects in their own departments.

  To demonstrate the top-down philosophy, the first eight projects were assigned to directors, who undertook to tackle some of the chronic problems that had been with the company for a number of years; this was a high-profile communications exercise. It was a deliberate policy to tackle improvement projects first and introduce quality circles once sufficient experience was gained. In this way quality circles were introduced into an environment where managers were already involved in quality improvement activities, thereby enhancing their chances of success.[41]

- *Phase two was measurement.* During phase one it became obvious to senior management that better methods of finding the root causes of quality failures were needed. An extensive training programme was undertaken in the use of SPC, multivariate analysis and Taguchi design of experiments. The Juran package[13] was adopted to provide a structured framework for this training.

- *Phase three was process management.* The first two phases resulted in a large number of people carrying out quality improvement activities. This led to the need to integrate a large number of projects and quality improvement teams into a coherent framework; process management was considered to be an ideal way to establish this.

  This necessitated looking at every activity in the company to analyse its purpose, its owner, and associated inputs and outputs in order to answer such questions as: 'What is the purpose of my department?', 'What are the non-value added activities?', 'Who are my customers and what do they expect of me?', 'How do they measure our performance?', 'What performance measures should be adopted?', 'What can I expect from my supplier?' In most cases, both the supplier and the customer were internal company departments. This phase required considerable effort, resources and inter-departmental co-operation in the form of a senior management working group.

  The key to this company's approach to quality improvement was top management's willingness to commit money and managerial resources over a relatively long time-span (three years), and to create an organisational culture and infrastructure to change the way the company did business.

**Company 3**

In contrast, the final case study provides an example of a low-key approach to quality improvement. The original starting-point was the need to implement SPC in response to the requirements of Ford's Q-101 Quality System Standard.[38] Because of money and resource restrictions, the quality improvement process was initiated within one production department at a time. Informal task-forces were used to identify and solve quality problems affecting both the manufacturing process and the product. SPC was introduced at the same time on a department-by-department basis. In-house SPC training courses were held for operators which acted as a forum for two-way communication between the quality department and operators, thereby enhancing the problem-solving process. FMEA was used on new products, and control plans were drawn up through the advanced quality planning process. The quality department also made use of the supplier quality assurance expertise of several of its major customers, in particular that of the Ford Motor Company, as a source of free advice on how to develop their quality system and planning methods.

Whilst it is tempting to cite this company as an example of a somewhat half-hearted top management commitment to quality improvement in delegating what it appeared to perceive as a 'bolt-on' technical task to the quality department, it is worth pointing out that considerable improvements were achieved, not least in customer satisfaction. The reason for this lies in the fact that the quality department developed an effective and realistic improvement strategy, making the best use of available resources, and kept to the task without losing sight of its objective. This shows how organisations can effect process improvement by keeping it simple and sticking to the basics.

## Summary

This chapter has focused on what we believe to be the main means of getting started on quality improvement – quality management tools and techniques, the teachings of quality management gurus, quality management consultancy, and quality system registration and customising – an approach based on these four methods in conjunction with appropriate received wisdom on the subject. The merits of each approach have been examined.

In our view, the means and approach used to start a process of improvement are of secondary importance compared to the creation of a clear strategic direction. The leadership displayed by members of the senior management team and their commitment to continuous improvement is critical to its successful introduction.

## References

1. Smith, S., 1986, *How to Take Part in the Quality Revolution: A Management Guide,* PA Management Consultants, London.
2. Dale, B. G., 1990, "TQM: What Are the Main Difficulties?" *The TQM Magazine*.
3. Dale, B. G. and Shaw, P., 1990, "Failure Mode and Effects Analysis in the Motor Industry: a State-of-the-Art Study," *Quality and Reliability Engineering International*, **6** (37), 179-188.
4. Shaw, P. and Dale, B. G., 1987, "Some Problems in Applying Statistical Process Control," *Quality Assurance*, **13** (1), 14-17.
5. Goldsmith, W. and Clutterbuck, D., 1984, *The Winning Streak*, Weidenfeld and Nicolson, London.
6. Peters, T. J. and Austin, N., 1985, *A Passion for Excellence*, Collins, London.
7. Ishikawa, K., 1976, *Guide to Quality Control*, Asian Productivity Organisation, Tokyo.
8. Lascelles, D. M. and Dale, B. G., 1988, "A Study of the Quality Management Methods Employed by UK Automotive Suppliers," *Quality and Reliability Engineering International*, **4** (4), 301-309.
9. Crosby, P. B., 1979, *Quality is Free*, McGraw-Hill, New York.
10. Deming, W. E., 1982, *Quality, Productivity and Competitive Position*, MIT Press, Cambridge, MA, USA.
11. Feigenbaum, A. V., 1983, *Total Quality Control,* McGraw-Hill, New York.
12. Juran, J. M., 1988, *Quality Control Handbook*, McGraw-Hill, New York.
13. Juran, J. M., 1979, *Quality Management Workbook*, Juran Enterprises, New York.
14. Dale, B. G., Lascelles, D. M. and Plunkett, J. J., "The Process of Total Quality Management," in Dale, B. G. and Plunkett, J. J., 1990, *Managing Quality*, Philip Allan, Hemel Hempstead.
15. Fine, C. H., 1985, *Managing Quality: a Comparative Assessment,* Booz Allen and Hamilton Inc., New York.
16. Main, J., 18 August 1986, "Under the Spell of the Quality Gurus," *Fortune*, 24-27.
17. Deming, W. E., 1986, *Out of the Crisis*, MIT Press, Cambridge, MA, USA.
18. Feigenbaum, A. V., 1956, "Total Quality Control," *Harvard Business Review*, **34** (6), 93-101.
19. Geeruliet, V., 1984, *Three of a Kind: A Reflection on the Approach to Quality*, Corporate Quality Bureau, Philips Group NV, Eindhoven.
20. Bendall, T., 1989, *The Quality Gurus*, Department of Trade and Industry, London.
21. McBryde, V. E., 1986, "In Today's Market, Quality is Best Focal Point for Upper Management," *Industrial Engineering*, **18** (7), 51-55.
22. Imai, M., 1986, *Kaizen: The Key to Japanese Competitive Success*, Random House, New York.
23. Ishikawa, K., 1985, *What is Total Quality Control? The Japanese Way*, Prentice-Hall, Englewood Cliffs, NJ, USA.
24. Mizuno, S., 1988, *Managing for Quality Improvement – the Seven New QC Tools*, Productivity Press, Cambridge, MA, USA.
25. Nemoto, M., 1987, *Total Quality Control for Management*, Prentice Hall, Englewood Cliffs, NJ, USA.
26. Shingo, S., 1986, *Zero Quality Control: Source Inspection and the Poka-Yoke System*, Productivity Press, Cambridge, MA, USA.
27. Suzaki, K., 1987, *The New Manufacturing Challenge: Techniques for Continuous Improvement*, Collier Macmillan, London.

28. Taguchi, G., 1986, *Introduction to Quality Engineering*, Asian Productivity Organisation, Tokyo.
29. Snowdon, M., 1986, "The Japanese Approach to Productivity and Quality: A European's View," *International Journal of Technology Management*, **1** (3/4), 411-424.
30. Dorsky, L. R., 1984, "Management Commitment to Japanese Apple Pie," *Quality Progress*, **17** (2), 14-18.
31. Lee, S. M. and Ebrahimpour, M., 1985, "An Analysis of Japanese Quality Control Systems," *Advanced Management Journal*, **50** (2), 24-31.
32. McMillan, C. J., 1982, "From Quality Control to Quality Management: Lessons from Japan," *Business Quarterly*, **47** (1), 31-40.
33. Schonberger, R. J., 1982, *Japanese Manufacturing Techniques*, Free Press, New York.
34. Trevor, M., 1986, "Quality Control: Learning from the Japanese," *Long Range Planning*, **19** (6), 46-53.
35. Ministry of Defence, 1973-1976, *Def-Stan 05-21, Quality Control System Requirements for Industry, Def-Stan 05-24, Inspection Requirements for Industry, Def-Stan 05-29, Basic Inspection Requirements for Industry*, Ministry of Defence, London.
36. Allied Quality Assurance Publications, 1972-1983, *AQAP 1 to 13, NATO Quality Control System Requirements for Industry*, Ministry of Defence, London.
37. Warner, F. E., 1977, *Standards and Specifications in the Engineering Industry*, NEDO, London.
38. *BS 5750/ISO 9000, Quality Systems*, 1987, British Standards Institution, London.
39. Ford Motor Company, 1990, *World-wide Quality System Standard Q-101*, Ford Motor Company, Plymouth, MI, USA.
40. Mone, J., Hibbert, B. and Dale, B. G., 1991, "Initial Samples and Quality Improvement: A Study," *Proceedings of the Sixth National Conference on Production Research*, Hatfield Polytechnic, Bell and Bain, 459-463.
41. Dale, B. G. and Lees, J., 1986, *The Development of Quality Circle Programmes*, The Training Agency, Sheffield.

# Chapter 5
# Leadership

## Introduction

In Chapter 3 we proposed that TQI is a form of organisational change that requires the intervention of a powerful champion – and that within the organisation the most powerful champion of change is the chief executive officer.

The view of internationally famous quality management experts such as Crosby,[1] Deming,[2] Feigenbaum,[3] Ishikawa[4] and Juran[5] is that unless the chief executive takes the lead to improve the performance of the organisation in relation to product and service quality, any quality improvement activities undertaken in isolation by individuals and/or departments will be only transient in nature.

In short, we are referring to leadership, and a special type of leadership at that: *transforming leadership.* This chapter reviews some of the issues and requirements in the leadership of TQI.

## The leader as a symbol

Numerous writers on the subject of TQM have stressed the crucial importance of the chief executive's role in the leadership of the quality improvement process. Mortiboys[6] says that: "... Quality improvement has become a vital cause for the chief executive to champion in his own company; a cause too important to be left to functional experts." The chief executive, through example and leadership, must influence others to become committed to continuous quality improvement. All the chief executives of the superior-performing companies with whom we have had contact have a vision for their organisation and, through personal leadership, have shared this vision with their workforces.

The concept of leadership has become firmly entrenched in the efforts of academic researchers and writers to understand and suggest improvements in the behaviour of organisations. The scale of the literature on the subject is staggering. Meindl and Ehrlich[7] report that the *Social Science Citation Index* reveals more than 3,000 entries under the single description 'leadership'; Stogdill's *Handbook of Leadership* contains over 5,000 references on the topic (Bass[8]). The concept of leadership has assumed a special status in the study of organisational behaviour, and in much the same way, so has the chief executive's role in the study of TQM.

On the other hand, there is a body of behavioural scientists who dispute the importance of leadership in the effectiveness and success of organisations, e.g., Aldrich[9] and Pfeffer and Salancik.[10] Their basic notion is that organisations and their managements are inherently dependent on the environments in which they operate. Such writers emphasise the process of attribution to explain behavioural causes, especially under ambiguous conditions. The following quotation from Pfeffer provides a useful summation of their argument:

> *There is a basic myth associated with leadership. We believe in attribution: when something happens we believe something has caused it. Leaders play that role in organisations. They take the credit for successes and the blame for losses. A more realistic conclusion would probably be that except in times of rapid growth, change or crisis, leaders don't make much of a difference in an organisation's performance.*[11]

Two points emerge from the above statement:

- First, the last two decades, at least, have been an era of rapid growth and decline, crisis and change, in which established behaviour, attitudes, norms and values have appeared under constant threat. Therefore one must question the validity of Pfeffer's last sentence. There are also a considerable number of well-publicised cases where the chief executive has changed an unsuccessful company into a successful one and vice-versa. Moreover, the process of quality improvement is an example of organisational change which requires effective leadership for it to happen and succeed.
- Secondly, an examination of some of the recent research on the process of attribution indicates that by acting as a focal point for perception, leaders can significantly affect an organisation's performance.

The research findings of Meindl and Ehrlich[7] conclude that people have a bias towards viewing leadership as a likely causal force when they account for organisational performance under ambiguous conditions, and that the strength of their preference co-varies with performance levels, i.e., leadership factors are credited for good performances and blamed for poor performances. They suggest that leadership is characterised by a powerful 'halo effect'. In short, leaders are potent symbols. Several empirical studies have reached a similar conclusion after analysing senior managements' attempts to exert influence symbolically, e.g., Staw et al[12] and Salancik and Meindl.[13] These studies speculate as to the reason why senior managers engage in acts of symbolism, e.g., to justify their own position, etc., but appear to miss the key point. The point is not whether a leader actually has the foresight, ability, judgement and skills to improve organisational performance, but whether the organisation's employees believe he/she has. For example, if the chief executive issues, on behalf of the senior management team, a company vision, mission or policy statement on TQM, it is symbolic act.

The role of the leader as a shaper of organisational values is closely allied to the concept of symbolic leadership. Burns[14] differentiates between transactional leadership and transforming leadership. He says that most leadership activities are transactional. Transactional leadership is a process of on-going transaction between a leader and his/her followers, whereas transforming leadership is said to build on people's need for meaning and to create institutional purpose. It should be noted that TQM is a transformational process. The transforming kind of leadership has a much greater strategic impact, because it involves real changes in attitudes, behaviour and norms. Research by Bass[15] revealed that charisma is the key factor in transforming leadership behaviour. He found that the subordinates to such leaders were inspired to have complete faith in the leader who was perceived as a role model to be followed – they believed in the symbol.

## Leadership and management

The terms 'leadership' and 'management' are usually used interchangeably, however they represent two fundamentally different activities. Kotter[16] summarises the difference: *management is about coping with complexity.* Management is concerned with the practises and procedures necessary to control and administer the business organisation which, as we argued in Chapter 2, is a bureaucracy in the strictest sense of the word. Effective management provides order and consistency to ensure the continuity of the complex environment that is the modern business organisation. By contrast, he goes on to argue that *leadership is about coping with change.* The need to cope with on-going change has become vitally important in recent years, for reasons such as faster technological change, greater international competition, over capacity in markets, national and international recessions, company mergers, the changing demographics of the workforce, the removal of European frontiers, political crisis, etc. More change always demands more effective leadership.

Kotter[16] argues that leadership is different from management. Leadership is not about having charisma or exotic personality traits; it is not the preserve of a chosen few. (In this respect Kotter tends to disagree with the views of Bass.[15]) Nor is leadership necessarily better than management, or a replacement for it. Leadership and management are complementary sets of activities necessary for success in an increasingly complex and volatile business environment.

Leadership and management are both concerned with three core tasks:

- Deciding what needs to be done.
- Creating networks of people and relationships to achieve these aims.
- Ensuring that what needs to be done is in fact carried-out.

However, these tasks can be approached in different ways.

Kotter[16] states that companies manage complexity by:

- Planning and budgeting – setting targets or goals, establishing detailed steps to achieve these targets, and then allocating resources to accomplish the plan.
- Organising and staffing – creating an organisational structure and sets of jobs for accomplishing the requirements of the plan, staffing the jobs with qualified people, communicating the plan to these people, delegating responsibility for executing the plan, and devising systems to monitor implementation.
- Controlling and problem-solving – monitoring results against the plan, identifying deviations, and planning and organising to solve the problems.

Whereas, the function of leadership is to bring about change by:

- Setting a direction – developing a vision of the future and the strategies needed to achieve that vision; describing a business, technology or corporate culture in terms of what is should become in the long-term and articulating a feasible way of reaching this goal.
- Aligning people – communicating the new direction to those who can create alliances that can understand the vision and are committed to its achievement (the fashionable name for this set of activities is 'top-down cascade').
- Motivating and inspiring – keeping people moving in the right direction, despite barriers to change, by appealing to basic but often untapped human needs, values and emotions.

It is clear from this that management is a deductive set of activities attempting to impose order on chaos, whereas leadership is an inductive set of activities attempting to bring about a new direction. Leadership is inspirational, breaking new ground; management is concerned with minimising risk and failure. Leadership is close to Burn's[14] concept of 'transforming leadership', whereas management is close to his concept of 'transactional leadership'. The purpose of management is to create systems and structures to help people successfully complete relatively routine jobs on a daily basis. It may be unexciting and unglamorous, but it is essential. Leadership can never be a substitute for management, and management can never be an excuse for leadership. Both sets of activities are inter-related and inter-dependent. They are part of a continuing cycle of setting a new course for the organisation and then keeping it on course in uncharted waters.

## Are leaders an elite?

Four propositions may be developed from the literature on the subject of leadership:

*Proposition 1*: more change always requires more effective leadership.
*Proposition 2*: the need for change is increasing.
*Proposition 3*: leadership is a growth industry.
*Proposition 4*: good leaders are born not made.

From this we might conclude that there is an acute shortage of good leaders which constrains an organisation's ability to change. This is an attractive, but flawed, logic flow because we believe Proposition 4 is incorrect. We are surrounded by people with the potential to become leaders, but who have not received the opportunity or nurturing to prove their leadership skills. Leadership is not, we believe, a set of personality traits, but a set of skills that can be taught. It is our view that the skills which leaders should possess include enthusiasm, knowledge, sound judgement, empathy of people, and effective communication. They must be prepared to win the trust and co-operation of the people they wish to lead – respect must be earned. It is obvious that not everyone has the inherent raw ability to perform the leadership function, in the same way that not everyone can become a good driver or teacher or politician, or general or football manager, but leadership is not the preserve of a gifted few destined for greatness. Any organisation able to unearth a naturally charismatic transforming leader will be fortunate indeed, particularly if that individual becomes, against the odds, chief executive. Organisations should seek to identify and develop a cadre of leaders from within their own ranks to become change agents, perhaps emulating the army recruitment motto: 'If you've got it in you we will bring it out'. In other words: *organisations should seek to develop a leadership culture.*

Most psychologists now believe that leadership is 'situationally determined'. The effectiveness of the leader, his/her potential for leadership, is dependent on the circumstances in which that person finds him/herself. A person might be a great leader in a certain set of circumstances but a disaster in another. TQI is about creating a corporate culture which is conducive to change; it is also about creating a leadership culture which enables people to act as change agents. This is not the same as creating a leadership elite which is preoccupied with restricting the distribution of power, and inevitably acts as a barrier to change. A leadership culture is concerned with extending the distribution of power throughout the organisation, with unearthing, training, developing and involving more leaders to empower networks of interaction within the organisation.

As Price[17] points out, leadership is a social exchange dependent on communication between the leader and followers. This cannot exist in a social vacuum – it is expressed in interactive situations. Leadership is a set of skills that can be learned – by experience.

Kotter[16] reveals that those companies which have consistently demonstrated the ability to develop leaders provide their potential leaders with plenty of leadership opportunities to learn and develop the appropriate skills. A significant leadership challenge early in a person's career is a key building-block: the opportunity to lead people to a shared vision, to take risks, build trust, delegate authority, and learn from success and failure, gives the trainee leader a perspective that cannot be taught in the classroom. The experience not only teaches how difficult the leadership task is, but gives an insight into its potential for generating change. It is clear then that leadership is about character formation and building, and the broader and more difficult the experience the better. Leadership development experiences should transcend the typically narrow experience of the traditional managerial career path or functional specialism – this is probably the main reason why middle management is so often a barrier to change. Broad-based leadership development is provided by a series of lateral career moves, or early promotions to broad job or special task force assignments, or a lengthy general management course. Not only does breadth of knowledge and experience assist the development of leadership skills, but the building of a network of relationships (both inside and outside the organisation) lays the foundation for lubricating future leadership initiatives. Japanese companies move their managers and technical specialists between functions on a regular basis. This not only helps to disseminate knowledge and technology, but also makes individuals more sympathetic to the needs, requirements and pressures of different types of functions.

To become effective leaders, managers must learn to perform the key tasks which differentiate leadership from management – setting a direction, aligning people, and motivating and inspiring.

It is not enough to accept a vision and declare a personal commitment. Leaders must be prepared to translate the vision into a new direction for the network of people for whom they are responsible. This means taking time to analyse information to make sense of the patterns of data, and to visualise what this means for the future of the network in terms of people and operational needs and its future relationship with other networks inside and outside the business.

Aligning people in the new direction is all about establishing a collective sense of purpose within the network. This is essentially a communications challenge – the sharing of a vision. To be an effective communicator, a leader must be prepared to be a team player, and this requires credibility. The leader must be perceived as a credible communication source if people are to believe in the message. In addition to the content of the message itself, the communicator must have a reputation for integrity and trustworthiness – or at least lack a reputation for the opposite. The communicator must also demonstrate a consistency between word and deed. This last element is easier said than done because it requires strong self-discipline by the leader. He/she must also check, from time to time, the adequacy of his/her communication to see if the delivered information has reached its source.

Motivation is about energising people. To quote Kotter:

> *Since change is the function of leadership, being able to generate highly energised behaviour is important for coping with the inevitable barriers to change. Just as direction-setting identifies an appropriate path for movement, and just as effective alignment gets people moving in that direction, successful motivation ensures that they will have the energy to overcome obstacles.*[18]

Motivation energises people by satisfying basic human needs such as achievement, belonging, recognition, self-esteem and control (Figure 5.1). Therefore, good leaders:

- Articulate their vision in terms which will address these intrinsic values, and will involve people in setting the new direction, giving them a sense of control and belonging, widening the scope of ownership so that an increasing number of people see the vision as theirs.
- Help people to grow, to enhance their self-esteem, by providing support through counselling, coaching, feedback and role-modelling.
- Recognise and reward success to give people a sense of achievement and reinforce the sense of belonging.
- Seek to replicate the leadership culture throughout the organisation, encouraging people to set new direction paths for their own part of the network.

The task of the leader is not to act as a reservoir of power, but to serve as a conduit of power to energise the people of an entire organisation. But chief executives and senior management teams should be aware of one crucial facet. The creation of a leadership culture is an exercise in power sharing – it really does amount to giving power to the people. They must ask themselves questions such as: 'Are we ready for this?' and 'Are we prepared for the consequences?' Although in an era of ever more change, the question becomes: 'Do we have any choice?'

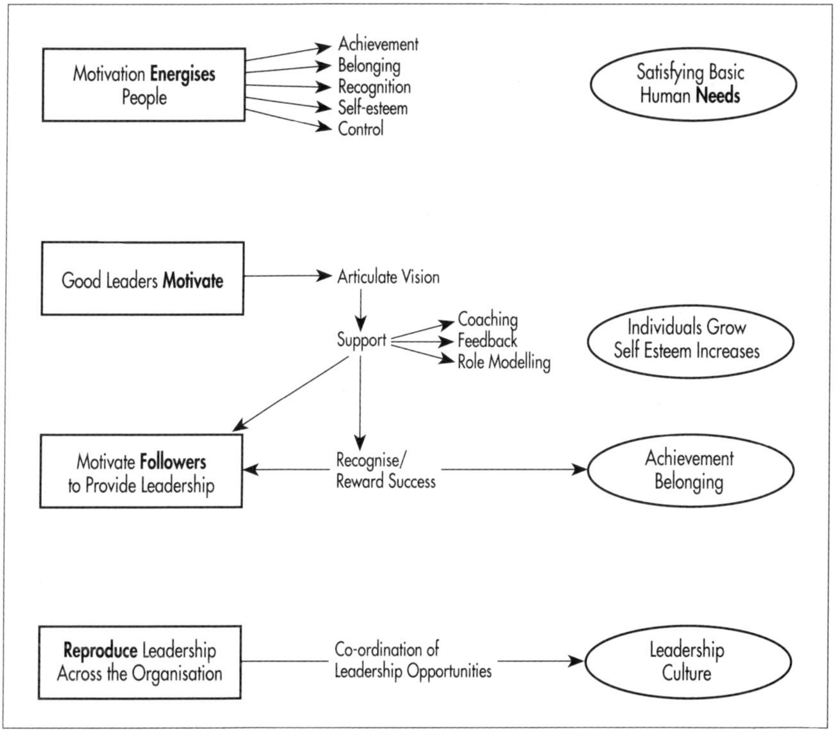


**Figure 5.1** Motivation as a leadership task

TQI needs leaders in all areas of a company's operation, and it is our view that organisations should consider, in some detail, the skills required for leadership in continuous improvement.

## Chief executives' perceptions of quality management

Recent independent surveys carried out by McKinsey[19] in Western Europe and Harbridge House[20] in the US revealed that the majority of chief executives recognise the importance of TQM to the competitive position of their companies.

In the McKinsey survey[19] of the chief executives of Western Europe's 500 largest companies, 87% of respondents stated that quality was one of their top priorities, and 91% recognised quality as 'absolutely critical' to the purchasing decision.

In a survey of 64 chief executives of US manufacturing companies conducted by Harbridge House, 77% stated intensifying pressure for TQM would change, substantially, future strategic or organisational priorities in their companies.[20] Moreover, as many as 92% of respondents stated that they would personally give greater attention to quality management in setting their company's direction over the next few years.

However, our own survey (carried out some two years earlier than the McKinsey and Harbridge House surveys) of the awareness and attitudes of 83 UK-based chief executives to TQM and their perceptions of TQM revealed that, whilst international competition and customer pressure may have sharpened chief executives' focus on the importance of product and service quality, their commitment is often limited.[21]

The main findings of our survey were:

1. *Quality was seen as an important competitive factor*
   Almost all the respondents believed that product and service quality was an important factor in international competitiveness. More than half had come to this conclusion within the past four years. Quality was ranked with price as the most important factor influencing sales. It was clear that whilst chief executives' attitudes towards quality improvement had become more positive, and awareness of the importance of product and service quality had increased, traditional perceptions of product and service quality still prevailed, e.g., quality could be inspected into the product or service, quality was the responsibility of the quality department.

2. *Customer pressure was the key trigger*
   Demanding customers appeared to be the most important agent for changing chief executives' behaviour, attitudes to quality, and their desire to motivate improvement initiatives and activities (see Chapter 3).

3. *There was a lack of a pro-active strategy for quality improvement*
   The motivation to improve quality appeared, in most cases, to be a reaction to demanding customers and ad hoc cost pressures, rather than part of a pro-active strategy to increase market-share, reduce costs, and increase profitability.

4. *Management by exception was the rule*
   Whilst the majority of chief executives recognised that they bore the ultimate responsibility for the management of quality im-

provement in their organisation, most had adopted a passive role which was limited to monitoring performance, in the words of one respondent, a 'watchdog' capacity. Many chief executives perceived TQM as a functional activity which could be delegated to a specialist, with their own role being limited to setting quality management policy and objectives and/or managing by exception. Less than one quarter of chief executives perceived themselves as standard bearers for quality, by, for example, taking an active and high profile role in attempting to raise quality awareness and develop an organisational culture conducive to TQM, chairing the TQM Steering Committee, leading quality improvement project teams and chairing quality improvement meetings.

5. *There was a lack of a corporate quality management policy*
   More than half the companies surveyed did not have a published quality management policy. Such a policy is an essential prerequisite for starting a formal process of quality improvement. Without a policy, improvement activities will lack cohesion, as different people in the organisation apply their own subjective interpretation of quality management policy, objectives and strategy. The findings revealed that few companies had corporate vision mission statements, and even fewer had in place any mechanisms for deployment of the quality policy through the organisation hierarchy.

6. *There was a low awareness of quality systems*
   A relatively large number of chief executives were unsure of the actual status of their company's quality system.

7. *Senior management had a limited knowledge of TQM*
   The majority of chief executives responding to the survey lacked knowledge of what was involved in the introduction and development of a process of continuous quality improvement.

8. *There was insufficient training in quality management*
   Personnel in less than half of the respondent companies had received training in quality management skills, tools and techniques. Most of those that had were quality staff, supervisors and work shop operatives. Relatively few directors and middle managers had received this form of training.

## The chief executive's role in quality improvement

From our research evidence we find that chief executives often say yes to the concept and principles of TQM and quality improvement, and then neatly side-step the issue by delegating its execution to the quality department, technical specialists, and/or other middle managers. Quality improvement requires the long-term commitment of the senior management team, with the chief executive taking a leadership role, but first he/she and his/her boardroom colleagues must understand the nature of TQM and perceive its relevance to the strategic vision and mission of the organisation. The majority of chief executives had recognised the need for quality improvement, and many had thought through the implications for the health of the business, but they still failed to translate this awareness into action. This is because TQM is still widely perceived in boardrooms as a collection of production-related tools and techniques delegated for implementation by the quality manager and technical specialists. In fact, TQM is a strategic multi-disciplinary process, which is nothing else but *total business improvement.* This fundamental fact has been understood by the superior-performing companies. TQM transcends departmental boundaries: everyone in the company manages quality, it is an operational responsibility. There is still a widespread failure amongst chief executive officers to appreciate these fundamental facts.

The chief executive's role is to ensure that each person in the organisation is committed to continuous quality improvement and that this process is cost-effective. This involves two key tasks: shaping organisational values, and establishing a managerial infrastructure to actually bring about change, i.e., an organisational culture that is conducive to continuous quality improvement. What is involved in these two tasks is now discussed.

### Shaping cultural values

The shaping of positive organisational values to create an environment which is conducive to lasting change and never-ending improvement depends on the nature of leadership within the organisation. The chief executive shapes the cultural environment of his/her organisation either through personal actions or by neglect. His/her actions may destroy the credibility of the entire quality improvement process within the organisation and also have a negative influence on the supplier community. For example, if he/she declares a commitment to quality but through personal actions emphasises that the schedule must be met at all costs, then he/she actively condones the production of non-conforming goods and/or delivery or non-conforming services – chief executives must 'walk the talk'.

The leadership style of the chief executive can act as a barrier to quality improvement; in many cases it is transactional (reacting to events) rather than transforming (shaping) future events (see Chapter 7). Unfortunately, this style of leadership does not readily lend itself to the kind of planned organisational change required for effective long-term improvement.

The chief executive's involvement in the quality improvement process is far more than a once-only input where he/she declares a commitment to quality, announces a quality policy and then delegates its implementation. Shaping organisational values requires real leadership, that is, leadership from the top – by setting new policy directions – and leadership from the front – by personal example. The chief executive must confirm top management's ownership of the quality improvement process, and avoid any suggestion of quality management being some kind of 'bolt on' technical activity.

In our experience, chief executive officers and senior management teams tend to grasp relatively quickly the concept of TQM and the principles of quality improvement, and its importance for the corporate health of their organisation, but it does take some time for them to change their attitudes and behaviour patterns. As a result, there is a danger that people at lower levels in the organisational hierarchy will receive mixed and confusing messages. It is important that senior managers and their reportees recognise this fact when planning for TQI.

**Establishing an improvement infrastructure**

Having diagnosed the need for change, and having defined the new strategy, the chief executive must turn his attention to establishing a managerial infrastructure to implement the change. In order to be effective and overcome the resistance to change usually encountered early in the transition process, this infrastructure must be company-wide in focus, and concentrate on developing multi-disciplinary improvement skills, cross-functional teams and an effective management information system.

**Continued commitment**

As a direct reaction to considerable market pressures, an increasing number of chief executives are now taking more interest in quality improvement. This interest encompasses such activities as:

- Attending TQM conferences and training courses.
- Holding discussions with management consultants and academics.
- Setting quality objectives.
- Developing strategy.
- Establishing a long-range management vision for the organisation and developing strategic quality planning.
- Allocating resources for quality-related training.
- Contracting a quality management consultancy to provide advice, guidance, and training.
- Launching a process of quality improvement.

It is important that the chief executive's involvement is not limited to this kind of input, otherwise the quality improvement initiative will be short lived. Quality improvement is an on-going process which can take at least ten years to become embedded into an organisation's culture.

Chief executives should allocate a set amount of time each week to quality improvement related activities. The chief executive officer at Milliken (one of the 1989 Malcolm Baldrige Award winners), for example, devotes more than half his time to the company's 'pursuit of excellence' process.

In order to shape organisational values they need, for example, to talk continuously about quality improvement to build up a common corporate vocabulary, and allocate adequate resources for quality planning activities. They should act as good models (housekeeping is an excellent example of this), visit customers and suppliers to report the progress being made on quality improvement, and develop positive measures of product and service quality as seen by the customer. In fulfilling their role of creating an organisational improvement infrastructure, they must, for example, be prepared to chair the TQM co-ordinating committee or Quality Council, lead or at least champion quality improvement teams, encourage teamwork, recognise the efforts of their people who are good team members, attend quality improvement meetings, and recruit good calibre people to take responsibility for quality in all areas of the business.

**Common failings**

Dale et al[22] devote a complete chapter of their book to the common failings of chief executive officers in relation to TQM. Their main findings are outlined under the following broad headings:

*Time* – The chief executive officer and other key organisational decision-makers are not prepared to devote the time to learn about TQM.

*Resources* – There is a failure to commit the right level of managerial resources to TQM.

*Teamwork is not practised* – The chief executive officer and his senior management team do not function as a team to develop improvement objectives and plans, and identify the means by which they can measure organisational improvement.

*Training* – Insufficient money is invested in TQM education and training. The chief executive officer and senior management team do not expose themselves to TQM training and even when they do, relatively few get involved in leading TQM training through a cascade approach.

*Understanding* – Chief executive officers and senior managers have insufficient understanding of the philosophy and logic underlying TQM or the associated tools, techniques, systems and procedures.

*Mistaken beliefs* – Some chief executive officers have picked up the wrong signals and are failing to follow the fundamental principles involved in a process of quality improvement.

*Out-put and cost are the main priorities* – The chief executive officer, in particular, if they are from a financial background, tend to be very concerned with internally derived numbers, as this is the main means by which they are measured and judged. This results in the placing of considerable emphasis on meeting production, output and cost targets, and rigid adherence to systems with product and service quality taking a lower priority.

*Statistical inadequacies* – The majority of chief executive officers do not understand process variation and capability.

## Leadership through Total Quality: three role models

We have stressed the importance of the chief executive's commitment and involvement in quality improvement; leadership style; and acting as a role model for the organisation. In drawing this chapter on leadership to an end it is worth examining the experiences of three celebrated role models: James Houghton, Chairman of Corning Glass; David Kearns, ex-Chairman of Xerox; and Karel Vinck, Chief Executive Officer of the Bekaert Group.

**James Houghton**

James Houghton of Corning Glass has made quality the primary responsibility of his leadership. He calls this providing 'unquestioned leadership', by which he means leading from the front. He insists that leaders have to earn the respect of their people 'out on the battle line', because in the end quality is about personal interactions with others.

According to Houghton, quality is not a little pool for wading and paddling in – it is an ocean. If you do not take the plunge and become totally immersed you cannot hope to coax a whole organisation to jump in. Leadership from the top requires the chief executive to get very wet. Houghton comments: "When you dive into the water of quality, a lot of people, in fact most people, will be standing on the beach waiting to see if you'll climb out because the water's too cold. Even if you stay in the water, they may not come in straight away."

When, late in 1983, Houghton announced that quality improvement was going to be a major initiative within Corning, most of his people were cynical. He had been chairman for just six months, and their reaction was, 'uh, huh, new chairman, new programme. All new chairmen have to have a new programme, right? But don't worry, this programme will last only six months or a year and then it'll be gone, the same way as all new programmes'. They were wrong; quality improvement did not go away. Since then, Houghton has not stopped talking about quality, concentrating on it, renewing his involvement – this really does take some commitment. Every year Houghton visits some fifty Corning sites world-wide – speaking on only one theme – quality. He believes that the cynicism is gone from his company for two reasons: first, the long-term commitment demonstrated by the top management team; and second, the acceptance that quality means the company's continued survival, not just the chief executive's 'flavour of the month'.

**David Kearns**

In April 1989, *Quality Progress* published a now famous interview with David Kearns, then Chairman of Xerox.[23] The cover of that issue records a brief summary of Kearns' perception of the chief executive's role in leading quality: "Quality is not banners. It's not speeches. It's hard work." When once asked why he went through all the 'hassle' of demonstrating commitment to quality instead of just telling people what he wanted them to do, his reply was: "We tried that. It didn't work."

Kearns believes that behaviour is what counts: what you do, the decisions you make, how you handle people. At Xerox the promotion of senior executives is linked to their level of quality involvement and initiative. As part of its quality improvement strategy, Xerox conducted a company-wide leadership audit of managerial attitudes, relationships with people, and employee perception of management behaviour. The audit was an important feature of Xerox's Leadership Through Quality process designed to change the corporate culture. One of the audit's most significant findings was employees' perception that the company still made appointments and promoted people with a short-term focus. Who gets promoted and recognised is one of the key signals of senior management behaviour. This led Kearns and his senior colleagues to introduce a new management assessment scheme based on quality. The scheme has three simple categories. The first level is that when talking about quality, in the use of tools, and in the depth of understanding, the manager is a role-model. The second level is where the individual can claim competence in quality management but is not yet a role-model. The third level is where the individual 'needs work'. If the statement 'needs work' appears on a manager's annual review report it affects his/her pay, bonus and eligibility for promotion. In fact, all the things that concern individuals the most. This is a powerful incentive for managers to become leaders and role-models.

Kearns has turned the need for a leaner and flatter organisation into a tool for empowering his leaders. In the late 1980s and early 1990s, growth for companies like Xerox has slowed down, meaning less people and fewer promotion opportunities. As a result, managers are staying in the same jobs for longer than before and learning to do things well. Instead of being mere batons to be passed on in a frenetic relay race to the boardroom, managerial jobs are actually becoming more meaningful. There is an incentive to make the most of it. Kearns believes that Total Quality is a form of empowerment that makes a managerial job more satisfying.

**Karel Vinck**

On becoming chief executive of Belgian steel products manufacturer, Bekaert Group, in 1985, Karel Vinck took the decision to make TQM a company-wide business strategy. It was for the significant and sustained improvement Bekaert achieved under his leadership that, in 1990, Vinck became the first recipient of the European Award for Leadership in TQM.

TQM was started in Bekaert as part of an overall competitive strategy for the Group. The top management first of all developed a beliefs and mission statement for the Group, and then looked for a vehicle to achieve their vision. They became convinced that TQM was the perfect vehicle. It became clear from the beginning that the top management team could not preach TQM and then by their own actions disregard it. So they decided to start by analysing and then improving their own performance. A special working group was established to improve the functioning of the board through the use of problem-solving tools.

Another challenge was how to communicate the TQM message across the organisation. It is hardly surprising that in a company over one hundred years old there was a reluctance to accept new ideas. That meant a sustained effort to change attitudes and the corporate culture. Initially, a few specific improvement projects were undertaken, leading to significant results which convinced the board that TQM should be systematically implemented through the whole Group. They realised that a lot of training would be needed – starting at the top. For example, every member of the top management team, including Vinck, had to be trained in leadership – a radical step for any top management team, even one with a declared commitment to change. Bekaert's improvement strategy succeeded because Karel Vinck and his top management team were prepared to take the time to improve their understanding of TQM and the methodologies involved; they accepted the heavy pull on their time and resources and were not too proud to expose themselves to training in the fundamentals. Vinck believes that leaders have to be the first to realise that they need training.

## Some leadership criteria

There are some very tangible criteria by which chief executives and senior managers, e.g., plant directors, general managers, divisional heads, etc., can measure their effectiveness as TQM leaders. For example, the European Quality Award for Leadership in TQM. As mentioned above, this was awarded for the first time in 1990 to Karel Vinck. The EFQM reported that his leadership inspired important improvements in the Bekaert Group's competitive performance and customer satisfaction.

The criteria for examination and judgement is divided into the following nine categories which all reflect essential elements of leadership in TQM:

1. *Drive* – Setting the scene for TQM.
   - Leads the TQM Steering Group and/or Council.
   - Develops a TQM model for the company.
   - Manages the company's change process.
   - Sets priorities on agendas.
   - Integrates all improvement activities into TQM.
   - Installs process management and process ownership.
   - Makes TQM visible.
   - Makes TQM a management responsibility.
   - Adapts appraisal and reward systems to support TQM.
   - Sets the process to involve all employees, customers and suppliers.
   - Develops a management role-model.

2. *Direct* – Providing business direction for the TQM process.
   - Participates in competitive benchmarking.
   - Integrates strategic planning/business planning and TQM.
   - Sets long-term objectives for Total Quality.
   - Clarifies short-term priorities.
   - Guides suppliers.

3. *Deploy* – Sharing business direction within the organisation.
   - Communicates on benchmarks, strategy, objectives and priorities.
   - Cascades objectives and priorities to all functions and levels in the organisation.
   - Checks employees' awareness of competitive position, strategy, objectives and priorities.
   - Invites participation in the deployment process from all employee, customers and suppliers.
   - Supervises consensus in the deployment process.

4. *Support* – Providing the necessary conditions for implementation.
   - Implements the structure for TQM.
   - Provides resources for TQM.
   - Approves support functions and staffs them.

- Facilitates involvement of employees, customers and suppliers.
- Sets, implements and maintains measurements.
- Organises the complaints handling system.
- Provides outside assistance if needed.

5. *Participate* – Committing personal involvement in the TQM process.
    - Performs visibly in TQM activities.
    - Leads education-and-training activities.
    - Takes part in improvement projects.
    - Accepts process ownership for vital processes.
    - Demonstrates personal improvement by attitude, behaviour and decisions.
    - Provides accessibility for employees, customers and suppliers.

6. *Review* – Monitoring the effectiveness of the TQM process.
    - Leads management audits, appraisals, assessments and surveys.
    - Visits customers to appreciate needs and progress.
    - Handles complaints from customers.
    - Visits suppliers to assess progress.
    - Evaluates progress in customer satisfaction.
    - Considers the effectiveness of Human Resources Management system for TQM.

7. *Recognise* – Motivating participants in the TQM process by recognising achievements.
    - Shows appreciation internally to groups and individuals.
    - Acknowledges performance of suppliers.
    - Appraises managers according to TQM principles and reacts to their achievements.

8. *Communicate* – Promoting all aspects of TQM, both externally and internally.
    - Attends to the long-term priority and consistency of TQM messages.
    - Makes progress visible both internally and externally.
    - Stresses values through communication.
    - Participates in external (national, branch, local) recognition activities.
    - Exchanges information on TQM with local, national and European authorities.

9. *Renew* – Continuously giving new stimuli to the TQM process.
   - Reads relevant literature and uses it for stimulation throughout the organisation.
   - Visits other organisations to study their TQM processes.
   - Makes TQM innovation a priority on the agenda of the steering committee.
   - Stimulates all employees to participate in continuous re-invigoration.

These criteria which are related to leadership achievements over a period of at least three years provide what we believe to be an excellent framework for self-assessment by the chief executive.

Examples of specific leadership achievements in each of the nine categories can be seen as milestones on the TQM Champion's road map – from setting the scene for the TQI process to stimulating its sustenance and renewal.

## Summary

This chapter has argued that the integration of the TQM concept and philosophy throughout all an organisation's functions and processes has to be management led. It was also pointed out that once senior management have demonstrated their commitment to quality improvement, then leaders and champions will emerge from all parts of the organisation.

The question of what constitutes leadership and management is not an easy one to answer, and there are many interpretations and views on the subject. However, an attempt has been made in this chapter to examine these related issues and point out to the reader the differences between management and leadership along with some of the key pieces of reading; the writings of Kotter[16,18] are worthy of special mention, and they have been quoted extensively in the section on leadership and management.

The perceptions which chief executive officers hold about TQM and their leadership role in quality improvement have been discussed along with the criteria, developed by the EFQM, by which executive management can judge their effectiveness as TQM leaders.

In closing this chapter on leadership it is worthwhile quoting the words of Protzman and Sarasohn in their discussion of democratic management: "A leader's main obligation is to secure the faith and respect of those under him."[24]

## References

1. Crosby, P. B., 1979, *Quality is Free*, McGraw-Hill, New York.
2. Deming, W. E., 1982, *Quality, Productivity and Competitive Position*, MIT Press, Cambridge, Massachusetts.
3. Feigenbaum, A. V., 1983, *Total Quality Control*, McGraw-Hill, New York.
4. Ishikawa, K., 1976, *Guide to Quality Control*, Asian Productivity Organisation, Tokyo.
5. Juran, J. M., 1988, *Quality Control Handbook*, McGraw-Hill, New York.
6. Mortiboys, R., 1984, "When Quality is Considered Why Buy British?", *International Journal of Quality and Reliability Management*, **1** (1), 15-19.
7. Meindl, J. R. and Ehrlich S. B., 1987, "The Romance of Leadership and the Evaluation of Organisation Performance," *Academy of Management Journal*, **30** (1), 91-109.
8. Bass, B. M., 1981, *Stogdill's Handbook of Leadership*, Free Press, New York.
9. Aldrich, H. E., 1979, *Organisations and Environments*, Prentice-Hall, Englewood Cliffs, NJ, USA.
10. Pfeffer, J., and Salancik, G. R., 1978, *The External Control of Organisations*, Harper and Row, New York.
11. Pfeffer, J., 1977, "The Ambiguity of Leadership," *Academy of Management Review*, **2** (1), 104-111.
12. Staw, B. M., McKenzie, P. I. and Puffer, S. M., 1983, "The Justification of Organisational Performance," *Administrative Science Quarterly*, **28**, 582-600.
13. Salancik, G. R. and Meindl, J. R., 1984, "Corporate Attributions as Strategic Illusions or Management Control," *Administrative Science Quarterly*, **29**, 238-254.
14. Burns, J. M., 1978, *Leadership*, Harper and Row, New York.
15. Bass, B. M., 1985, *Leadership and Performance Beyond Expectations*, Free Press, New York.
16. Kotter, J. P., 1990, *A Force for Change: How Leadership Differs from Management*, Free Press, New York.
17. Price, F., 1990, *Right Every Time*, Gower, Aldershot.
18. Kotter, J. P., 1990, "What Leaders Really Do," *Harvard Business Review*, May-June, 103-111.
19. McKinsey and Company, 1989, *Management of Quality: The Single Most Important Challenge for Europe*, European Foundation for Quality Management, 19 October, Eindhoven.
20. Anon, 1990, *Managing Complex Change, A Harbridge House Top Management Survey*, Harbridge House Inc., Boston, Massachusetts.
21. Lascelles, D. M., and Dale, B. G., 1990, "Quality Management: the Chief Executive's Perception and Role," *European Management Journal*, **8** (1), 67-75.
22. Dale, B. G., Cooper, C. L., and Shaw P., 1992, *Total Quality and Human Resources: An Executive Guide*, Blackwell Publishers.
23. Anon, 1989, "Payment in Kind (an interview with David T. Kearns)," *Quality Progress*, April, 16-20.
24. Wood, R. C., 1989, "A Lesson Learned and a Lesson Forgotten," *Forbes*, 6 February, 70-78.

# Chapter 6
# Supply-Chain Management

## Introduction

The quality of purchased supplies is critical to the quality of a company's product and/or services. Crosby estimates that 50% of a company's quality problems are caused by defective purchased materials.[1]

Moreover, for many companies, the purchase of materials from suppliers and sub-contractors accounts for at least 50% of manufacturing costs, and in some cases, e.g., the automotive industry, this figure can be over 70%. It is clear that many organisations rely heavily on their suppliers as a major resource. So, any company engaged in a process of continuous improvement needs to seek improvements in quality and productivity from its suppliers, and to integrate them into its own quality improvement process.

There are two other main reasons why supply-chain management and supplier development have become key elements in improving a company's competitiveness.

First, technological and competitive pressures have resulted in a greater trend towards specialisation – an increasing number of organisations developing strategies which concentrate on their core businesses. The fact that few companies can maintain, in-house, the complete range of expertise needed to keep pace with the latest advances in technology (in manufacturing processes as well as products), coupled with increasingly global competition, is forcing vendors to develop specialised capabilities.

As specialisation increases, so does the number of parts sourced by an organisation from its external suppliers; there is a general trend from making to buying. Viewed strategically, however, decisions on suppliers' roles go beyond simple choices about what to make in-house and what to source externally; the need to involve suppliers early in the process of new product development, in order to obtain their inputs and make use of their specialised expertise, becomes of prime importance. This early involvement and the two-way flow of information facilitates design for manufacturability, simultaneous engineering and competitive manufacturing.

Second, the nature of competition itself has changed. Feigenbaum,[2] among others, postulates that today effective international competition is a combination of competition in its traditional and visible form (product versus product and/or service versus service) and an equally powerful, but less visible, form of competition involving companies' skill in implementing and managing a total approach to quality improvement. Suppliers are a vital part of this. Therefore, those companies with the best suppliers, and which can make the most effective use of their capabilities, will have a competitive advantage in the market-place. Supplier development requires a fundamental shift in the customer-supplier relationship; it requires a company to treat its suppliers as long-term business partners, and this is a strategic decision.

It is also worth noting that Debruicker and Summe[3] and Roberts[4] point to the fact that suppliers should be more proactive in developing their customer base as a competitive strategy, instead of merely reacting to the quality improvement demands of their customers. It is our experience that suppliers approaching their customers in this way are considered by them to be forward looking and professional. This presents a unique marketing opportunity.

This chapter examines some of the current issues in supply-chain management and supplier development. It is written from our research and experience in the manufacturing industry, but the concepts explored are also applicable to those organisations providing 'services'. The supply of 'goods' whether in the form of material, parts, data, systems and services which accompany the provision of services are a key factor in determining service quality. Therefore it is important that service organisations work closely with their suppliers.

The chapter opens by arguing the need for a long-term collaborative business partnership between a customer and supplier before it goes on to explore the barriers to developing suppliers. The issues involved in starting supplier development are then explored along with the suppliers' development process.

## Long-term business partners

The traditional approach to customer-supplier dealings has been based on the assumption that they are adversaries – the customer and suppliers having differing objectives, and being engaged in a win-lose contest. Not surprisingly, this form of relationship tends to focus on negative issues and is often characterised by uncertainty. Communication with suppliers is guarded with as little information as possible being transmitted between the two parties, just in case one of them may turn it into an advantage at the expense of the other. In this situation, the customer's best protection was thought to be the establishment of several alternative sources of supply for each bought-out item, and regular switches from one supplier to another through the use of price-driven contracts, resulting in a large supplier base and limited co-operation between parties. The competitive environment was thought helpful to keep suppliers on their toes. But the end result was supplier reaction characterised by the comment 'you can have quality, but it will cost you'. Some people have likened the relationship (if one can call it that) to a game of cat and mouse.

The usual outcome is mistrust, fear, dishonesty and frustration in customer and supplier organisations which prevents long-term quality improvement taking place.

Clearly, supplier development requires a radically new form of customer-supplier relationship. The Philips Group[5] coined the phrase 'co-makership' to describe it. Co-makership means working together towards a common goal; it is based on the principle that both parties can gain more benefit through co-operation than by separately pursuing their own self-interests – a win-win situation. Co-makership means establishing a long-term business partnership between customer and supplier based on common aims and aspirations, mutual trust, co-operation, honesty, open declaration of problems, a willingness to work together to find answers to problems, clearly understood responsibilities, and a desire by customer and supplier organisations to improve the product continuously. As part of this trust, many relationships use the 'open book' method in arriving at the price per piece. The supplier shows to the customer his manufacturing and purchasing costs, costs of non-conformance, profit and loss account, etc. This analysis, which is carried out jointly by customer and supplier, identifies areas for improvement and provides some indication of the cost which can be taken out of the product.

Typical features of the approach that is being taken by some organisations to supplier development are:

- Long-term purchasing contracts (the condition which is usually attached to such a contract is for the supplier to agree to reduce the price of the component by a certain percentage each year).

- A willingness amongst both parties to learn more about each other's business, visibility of potential business and preferential opportunities to secure that business.
- Keeping the supplier informed of changing requirements.
- Providing suppliers with early access to a customer's future design and manufacturing plans.
- Suppliers making a contribution to the customer's design process.
- Supplier undertaking development work and monitoring technological trends.
- Integrating business practice.
- Joint problem-solving activities.
- Single sourcing.
- Reducing the supplier base.
- Reduction, and then elimination, in the inspection of supplied parts.
- The customer providing a consultancy and support service to suppliers.
- A comprehensive customer-supplier communications network using Electronic Data Interchange (EDI).

A good example of the benefits achieved from joint problem-solving activities between customer and supplier is provided by Dale and Plunkett[6] from a case study made of quality costing at the Leyland Daf Assembly Plant (LAP).

> *It was established that a German supplier was sending, each month, to LAP some £10,000 worth of non-conforming material (for example damaged parts, under-sized dimensions and holes missing). A Quality Improvement Team (QIT) was formed with the supplier and LAP personnel. This team first identified and analysed the non-conformities, from this they developed an understanding of the reasons why the non-conformities had occurred. Counter-measures were then put into place. These solutions often involved the application of short-term fixes followed by long-term permanent corrective action. By this means the supplier has been able to reduce non-conforming material to a monthly figure of £1,000. As a direct result of these successful quality improvement efforts, the supplier has offered to keep his prices to LAP at the 1988 level for two years. This form of quality improvement activity is carried out with all key suppliers of LAP.*

To develop this new kind of relationship, considerable changes in behaviour and attitude are required in both the customer and supplier organisations. Customers need to be prepared to develop plans and procedures for working with suppliers, and to commit resources to this. On the other hand, suppliers have to accept full responsibility for the quality of their shipped product, and not rely on the customers receiving inspection to verify that the product is to specification. As a prerequisite of the new relationship, both parties have to reach an agreement on how they will work together, i.e., establish the ground rules.

The Ford Motor Company is an example of a major organisation which has set out to build long-term partnerships with its suppliers in which a joint approach is stressed. 'It is Ford and her suppliers who make cars' is a philosophy which has existed since 1982. The following points illustrate how Ford of Europe endeavours to co-operate with its supplier community:

- A change has been made from Supplier Quality Assurance to Supplier Quality Assistance, with the introduction of a new combined function of Supplier Quality Engineering.
- Participation by suppliers in Continuous Improvement Teams, e.g., Quality Improvement, Quality Function Deployment, and Design of Experiments.
- Provision of training in SPC and other techniques to suppliers has been facilitated by the establishment of external Regional Training Centres which are approved and supported by Ford. This has been taken up by more than 6,000 employees from over 1,200 suppliers.
- Launch of the Interactive Video Programmes in SPC and FMEA as an on-going training aid. The SPC programme, developed in conjunction with Laser Media,[7] has won number of prestigious training awards.
- Introduction of the Q1 award in February 1988. From 1st January, 1992, no new business is awarded to non Q1 suppliers. To maintain carry-over business, suppliers must achieve Q1 status by 1st August 1993. Suppliers with a Q1 status have a preferred position in new product development and source selection.
- Reduction of the number of suppliers from 2,100 in 1980 to currently less than 1,000. This is to facilitate enhanced co-operation and improved quality.

- Increase in the number of long-term contracts of three-to-five years which have been awarded.
- Early vendor involvement and source nomination for new programmes, and improved communication between prototype and production buyers and suppliers.
- Suppliers are involved in a Just-in-Time and supplier communications programme.

A few researchers offer evidence which demonstrates the advantages of developing a co-makership type relationship with suppliers. For example, a comparative study of two electronics manufacturers by Masson[8] revealed that the manufacturer which had developed a 'co-makership' type of relationship achieved a significantly better performance from its suppliers than had the manufacturer which continued to operate the traditional adversarial approach. The benefits to the manufacturer which had adopted a 'co-makership' approach were found to be:

- Shorter delivery lead times.
- Reliable delivery promises.
- Less schedule disruption.
- Lower stock levels.
- Faster implementation of design changes.
- Fewer non-conforming items.
- Stable prices.
- A higher priority being given to their orders.

## Barriers to supplier development

The research we have carried out reveals that certain aspects of the customer-supplier relationship can act as a barrier to supplier development. These include:

- Poor communication and feedback.
- Supplier complacency.
- Misguided supplier improvement objectives.
- The credibility of the customer as viewed by their suppliers.
- Misconceptions regarding purchasing power.

These barriers are now examined.

**Poor communication and feedback**

In general, communication and feedback between customer and supplier is not good, moreover, both parties sometimes do not realise how poor they are at communicating with each other. The majority of suppliers with whom we have carried out research perceive the quality performance requirements of their major purchasers as realistic, however, a substantial number of them feel that communications and feedback between customer and suppliers could be improved. Most of the dissatisfaction appears to relate to technical specifications and product acceptance criteria, particularly the lack of consultation on design and product engineering aspects. A number of suppliers are also critical of their customers in relation to the frequency of changes to the delivery schedule. The three areas of major concern are:

- Performance feedback to suppliers.
- On-going communications with suppliers.
- Precise materials specification data.

There are some strong indications that not all dissatisfied suppliers actually communicate their dissatisfaction to the customer; a typical outcome of what may be regarded as an adversarial relationship. Despite this, most suppliers are happy to collaborate more closely with their customers on component design and product engineering activities. Such collaboration forms a solid basis for long-term improvement and provides a logical starting point for a supplier development activity.

Non-conformance of purchased items is often due to the customer's inability to communicate his/her requirements clearly. Writers such as Deming[9] and Ishikawa[10] are agreed as to the culpability of purchasing organisations in allowing this to happen. Ishikawa claimed that at least 70% of the blame for non-conforming purchased items lies with the customer. Derose[11] notes that many buyers have recognised this and are experimenting with various new approaches such as supplier base reduction, single or limited sourcing arrangements, and tighter integration with supplier planning and scheduling systems. It is up to the purchasing manager to ensure the existence of a clear specification which defines the exact requirements, but this in itself is not enough to assure conformance. The supplier must be given the opportunity to understand the function of the part being supplied and discuss design details, particularly with regard to key features and the manufacturability of purchased or sub-contracted items, before requirements are finalised.

The following three examples illustrate the problems caused by a lack of communication:

1. One manufacturer of automotive components had experienced frequent problems with the performance of its castings suppliers. A detailed investigation of the causes of dimensional and metallurgical non-conformances revealed that many of the problems were of the company's own making, due to its failure to specify requirements in a clear enough manner, and in not recognising the unique operating problems that existed in the foundry industry. The company discovered that the quality performance of its suppliers improved when suppliers with the appropriate expertise were selected, and the design and functional requirements of the finished casting were discussed with foundry representatives before the contract was signed.
2. The senior quality engineer of another firm had no doubt that communicating quality requirements to suppliers clearly and accurately was the biggest single breakthrough his organisation could make in improving supplier quality performance. In his firm, technical engineering standards were not always sent to suppliers with the purchase order. The purchasing manager, not surprisingly, disagreed, saying that most suppliers had been serving the firm for some time and should be well aware of the company's quality conformance requirements. He contended that the engineer's observation was in fact just an excuse used by some suppliers to ship non-conforming items. But like many of his colleagues from the traditional school of purchasing, he perceived all supplier feedback to be either critical or an excuse for poor performance. In addition, the internal information flow was inadequate, particularly with regard to the communication of technical engineering standards; the firm tended to function on a need-to-know basis.
3. Inadequate communications and lack of attention to initiating permanent corrective action was the cause of a heated discussion we overheard in one relatively large purchasing organisation. A problem with non-conforming holes had been found in a batch of fabricated parts, and the supplier was asked if they remembered that the same problem had occurred six months or so before, when the request was made that they bring their drilling equipment in in order to rectify the problem. The supplier had not taken the corrective action requested – they had forgotten.

Some purchasing managers and supplier quality assurance engineers seem to think that the *quality performance of their suppliers can be achieved almost by remote control.* Thus, suppliers are not visited and assessed on a regular basis and are not invited to discuss improvement strategies, plans and initiatives. Such people are disappointed and often surprised when non-conforming items are supplied. During the course of our research we have come across several people from major purchasers who genuinely believed that they engaged in joint quality-planning with suppliers, whereas in fact the communication process was all one-way (from customer to supplier) and feed-back from suppliers was discouraged (either because it was ad hoc and ignored, or because it was never sought in the first place). In such cases, quality planning is rarely joint, as suppliers have no say in the product specification or design. After the initial vendor approval process (which may have occurred several years previously) some purchasing organisations do not even bother to assess the current state of the capability of their suppliers' processes and quality systems, or to measure the suppliers' quality system. Indeed, in carrying out an assessment there is sometimes a tendency to concentrate on auditing the effectiveness of the paperwork, whereas the emphasis should be on assessing and encouraging improvement. There is a considerable reliance on quality manuals as evidence that a quality system is in place.

The following three examples of non-conforming materials occurred in one organisation we studied, and help to highlight the problems caused by inadequate communication and feedback systems. (The firm's annual purchasing expenditure is approximately £60 million, and covers over 200 different components and parts.)

1. The company experienced a high rate of in-process rejections on a particular type of purchased forging. The cause was eventually traced to the inability of the supplier's heat treatment process to adequately treat the bottom of the grooves cut into the component. Due to a design flaw, this was a customer requirement that eventually turned out to be impossible to fulfil by this or any other vendor. However, the supplier did not inform the buyer that the process could not meet the specification in full.
2. An identification code was required to be stamped on the outside rim of the head of a particular type of screw. The supplier continued to ship screws with the code stamped on top of the head. When questioned by the quality engineer about this non-conformance,

the supplier replied, "Why are you complaining now? We have supplied screws coded in this manner for years and you never complained before. Besides, we don't have a machine capable of stamping the outside rim."

3. A component was required to be manufactured from a specific grade of steel. On receipt from the supplier, the component material was found to be too soft for final machining. Subsequent investigation uncovered an interesting situation. The specified steel could be obtained only from a source in the West Germany in 50 tonne lots, a quantity that far exceeded the supplier's annual production requirements. Therefore, the supplier decided to manufacture the components from the 'next best' grade of material which was more easily obtainable.

These examples of non-conforming items occurred in a company that prides itself on having an efficient purchasing function, a lean and mean attitude towards business, a relatively small supplier base, practising Just-in-Time purchasing, and encouraging its suppliers to adopt SPC. This firm is a prime example of managing supplier performance by remote control. Despite its belief to the contrary, the firm does not engage in joint quality planning with its suppliers. There are a number of reasons for this:

- Strategically, the company is cost-driven. The management team is keenly aware of the fact that it is part of an American-owned multinational group operating in price sensitive markets. Moreover, the entire output of the plant studied goes to the US. Consequently, a key company objective is to keep material prices at not less than 30% below comparable materials used in its North American operations. One outcome of this sensitivity to cost is that the company seems to manage its supplier base as though it were involved in a commodity trading market. It also actively pursues an international sourcing policy.
- Long-term supplier involvement is not a noticeable feature of the firm's purchasing operation, and there are frequent switches from one supplier to another. For example, in one year alone, 37 vendors were removed from the company's approved list (almost all for price-related reasons) and 59 new suppliers added. The quality department's involvement in supplier selection tends to be ad hoc, and price rather than quality criteria appear to be the basis for selection.

- The company's management is constantly involved in fire-fighting activities, perhaps because of a 'macho' self-image.
- Its communications and feedback systems, both internally and with suppliers, are ineffective.
- The firm's management has an apparent inability to learn from past events and to put into place long-term corrective action, and the same non-conformances tend to crop up repeatedly.

**Supplier complacency**

Many suppliers appear to be complacent about customer satisfaction with the quality of their product or service. This situation was probably best summed-up by the following comment by Sir John Egan, the former Chairman and Chief Executive Officer of Jaguar Cars:

> *When in 1980, Jaguar started to take to task its component suppliers for poor performance of their product, they (component suppliers) were often surprised, as, until that time no-one had bothered to give them feedback of this kind and they, therefore, could be partly forgiven for believing that everything in the garden was rosy. I say partly because component suppliers did very little to find out how their products performed in service.*[12]

There are two types of measurement relating to a customer's satisfaction with the quality of a supplier's product and services; reactive and pro-active.

Examples of *reactive measures* include:

- Internal failure data, e.g., scrap reports, non-conformity analysis.
- External failure data, e.g., customer rejections, warranty claims.
- Customer assessment rating and audit reports.
- Verbal feedback from meetings with customers.
- Contractual requirements outlined in the customers' vendor improvement plans.

Examples of *pro-active measures* are:

- Customer interviews.
- Customer focus workshops and clinics.
- Surveys of customer needs.
- Evaluating competitors' products.
- Competitive benchmarking.
- Reliability analysis.

- Advanced quality planning carried out in conjunction with customers.

A large number of the organisations with whom we conducted research *failed to have adequate quantitative and pro-active measures*, e.g., a customer satisfaction index, of how well they were satisfying their customers' needs and expectations. In the main, the measures were reactive and centred on internal and external failure data. A number of suppliers saw customer satisfaction in very simple terms. For example:

- "If the customer does not return our product then the quality and reliability must be satisfactory."
- "Continual acceptance of delivered product."
- "If the customer does not complain they must be satisfied with our products, business system and service."
- Monitoring warranty claims was also claimed by some suppliers as a typical measure of customer satisfaction. One supplier said they monitored warranty claims to "identify the need for customer education or product enhancement."
- Adopting a specific quality management tool or technique in order to satisfy a customer is seen by some suppliers as a positive measure of customer satisfaction. One supplier stated, "Procedures of ever-developing statistical methods have reduced problems at our customers leading to repeat and new orders." It is noticeable that automotive companies appear to be this company's most important customer; hence the implementation of SPC. However, this company does not use any formal market research, so clearly its approach to customer satisfaction is reactive and not necessarily planned in advance as part of a marketing strategy; this is in spite of claiming that it has a total approach to the management of quality. They also rank general company reputation as the most important factor influencing the sales of its product – perhaps the adoption of SPC has enhanced its reputation amongst its customers.
- Another company has a very tangible measure of customer satisfaction, an IBM Quality Award. IBM is obviously a major customer and the company is sensitive to IBM's requirements, particularly with regard to JIT purchasing which is one element of IBM's supplier improvement process. Accordingly, this supplier ranks delivery as the prime factor influencing sales, followed

by reliability and general company reputation. Interestingly, performance in use (fitness for purpose) is not considered to be an influential factor. This company does not use any of the planning elements of quality improvement – its activities appear to be purely reactive in terms of customer satisfaction.

The direction of an organisation's focus on customer satisfaction appears to influence the use of specific quality management techniques. Our research reveals that companies which recorded internal failure data or lack of it as their prime measure of customer satisfaction were less likely to evaluate competitors' products. More 'outward looking' companies which used formal assessments by their customers as their prime measure were more likely to engage in market research and hold design review meetings.

It is clear that a considerable number of suppliers make no real efforts to find out what their customers think; an attitude of 'no news is good news' still appears to be the norm. This is a short-term view which will ultimately result in lost business opportunities.

**Misguided supplier improvement objectives**

Companies are often not sure what they want from their supplier improvement initiatives. Comments made by suppliers indicate that some of their customers do not understand the fundamentals of TQM. Many have formal vendor audit programmes but no clear supplier development objectives. There also appears to be a dilution and distortion of the quality message as requirements are passed down the supply-chain. For example, when faced with demands from customers for improved quality, suppliers usually react by implementing specific quality management tools and techniques, and in turn insist that their own suppliers use the same. Not all customers are actively involved with their suppliers in helping them to solve quality problems. Also at one time there appeared to be a blind faith in the power of SPC to do the trick, and it was clear that many companies assumed that introducing SPC was the same as developing a TQM approach. Fortunately this situation is changing as organisations have acquired operating experience of SPC.

Similarly, these companies are under the equally mistaken impression that the imposition of a particular quality management technique on their suppliers as a condition of purchase is the same as supplier development. But, rather than stimulate changes in behaviour and attitudes, the outcome tends to result in what Schon[13] calls 'dynamic conservatism', in that the imposition is seen as a disruptive threat to the organisation. If an organisation cannot neutralise the threat it will comply to the least possible extent with the requirements for change, i.e., 'dynamic conservatism'. It is very clear that the implementation of techniques without behaviour and attitude change means therefore that any benefits gained will only be short-lived (see Chapter 4).

**Lack of customer credibility**

A purchasing organisation's lack of credibility in the eyes of its suppliers is another barrier to supplier development. *Suppliers need to be convinced that a customer is serious about continuous quality improvement*, and this is conditioned by the customer's behaviour and attitudes. The following are examples of how a credibility gap may emerge.

- Poor purchasing and supplies management practises such as a competitive pricing policy, frequent switches from one supplier to another, unpredictable and inflated production schedules, last minute changes to schedules, poor engineering design/production/supplier liaison, inconsistent decisions made by SQA personnel, over-stringent specifications, abuse of power by SQA personnel and the threat of loss of business as a bargaining ploy all lead to a credibility gap in the customer/supplier relationship. It is not uncommon for a customer to talk quality to its suppliers and then act quite differently by relegating quality to secondary importance behind, for example, price or meeting the production schedule.
- Similarly, there is little value in holding quality improvement conferences and seminars for suppliers if the purchasing organisation continues to adopt an adversarial approach to its suppliers, or is seen to accord a low priority to quality unless there is a serious non-conformance – in the words of one supplier quality assurance engineer: "No one cares about vendor performance until the production line stops."
- The TQM image being created by major purchasing organisations in discussions with suppliers is not reflected in reality when supplier personnel visit their own manufacturing sites.
- Another way in which credibility is lost is when a customer has continued to accept non-conforming items over a long period of time, even if unwittingly, and then criticises the supplier for supplying non-conforming materials. In such cases the supplier

sees these newly imposed standards of conformance as somewhat arbitrary, and will tend to question the adequacy of the purchaser's quality system, standards and procedures in detecting non-conforming product.

- A supplier may report to the customer that the customer-supplied tooling is reaching the end of its useful life. The customer then asks the supplier to carry out some minor refurbishment as a short-term measure; the supplier advises against this but is pressured to do the repairs. When non-conforming parts are found in batches from the 'patched-up' tooling, the supplier is taken to task by the customer and acquires quality performance demerits. There is little doubt that major customers need to improve their methods and strategies for dealing with the tooling used for supplied parts.
- Failure to respond to a supplier's request for information, advice on queries, or feedback on specification requirements, component functionality, and to provide a design FMEA, etc.
- Approval of initial samples gives confidence to the customer that purchased products are to the specification at commencement of volume production. We have encountered the situation that a supplier's components are used by the customer without them passing the initial sample procedure. This happens when a supplier fails to comply with the procedure set down by the customer and the customer does not correct this shortcoming. All our evidence indicates that a high level of quality non-conformance is found on products not having initial sample approval.
- When the customer's SQA personnel are fooled by the camouflage measures, fakes and ruses employed by a supplier in an assessment of a vendor's quality system. This often applies in the case of SPC.
- A number of suppliers are cynical of JIT as operated by major purchasers. They often gain the impression that they are being forced to hold stocks for their customers. Typical of the comments we have heard are: "None of our customers really employ JIT, they just pretend that they do.", and: "It is only our Japanese customers who employ JIT purchasing correctly."
- Perhaps the most telling example of a lack of customer credibility is the following scenario: A batch of non-conforming parts is returned to a supplier, who does not scrap or rework the batch, but based on his experience of customer behaviour, returns it to his finished goods store. Some days later a telephone call is

received from the customer who exclaims that the production line is stopped awaiting the batch of parts. The untouched batch is then re-shipped to the customer who accepts it and feeds the parts onto the production line without any checks.

**Purchasing power: a misconception**

Purchasing power is a major issue in the buyer-supplier relationship. Lack of purchasing power is a commonly cited reason for lack of success in improving supplier quality performance. The general view is that a purchaser's influence on its suppliers varies with its purchasing power, and that the greater this power, the more effective its supplier quality assurance activities will be. During the course of our research, we came across cases that support this notion, for example:

- It is clear that major automotive companies and large public sector purchasing organisations have used their purchasing muscle to force their suppliers to adopt certain quality management techniques and practises, as a contractual condition of business. On the other hand, one firm (a British Quality Award winner) had little success in changing managerial behaviour and attitudes towards quality improvement in one of its large raw materials suppliers, due to its relative lack of purchasing power, even though it offered direct assistance to solve non-conformance problems.
- Another British Quality Award winner, presumably, therefore, a company with demonstrable quality standards, performance and an established reputation for TQM, still appeared to lack influence over its suppliers, even after having reduced its supplier base by almost 40%. Of the 20 places on a Department of Trade and Industry sponsored quality activity programme offered to suppliers at the company's annual quality convention, only seven were taken up, whereas all 40 places on offer to automotive component suppliers at the Society of Motor Manufacturers & Traders' Drive for Total Quality Conference were filled. The purchasing power of the major automotive companies was a significant factor.

Purchasing power alone is not a sure fire guarantee of improving supplier performance. Companies with considerable purchasing power may well improve the quality of purchased items, but will not necessarily achieve lasting benefits, or motivate their suppliers to internalise the benefits of a process of continuous quality improvement to satisfy all their customers. There is a tendency for some vendors to *treat powerful customers as sacred cows, leading to stratified quality assurance.* That is to say, companies grade the quality of their products at different levels according to past experience of individual customer expectations – 'Company A won't accept this nonconformity but Company B will'. In the early days of operation at a major Japanese-owned company in the UK, on more than one occasion the production line was in danger of being stopped because of failure of suppliers to deliver on time. The purchasing department discovered that one supplier was so in awe of the Japanese company's reputation that it had decided that component batches with minor surface imperfections could not be shipped, even though the imperfection was not detrimental to the functionality of the component.

The stratified quality assurance approach often stems from the traditional misconception (mentioned in Chapter 1) that quality is an optional product attribute or extra, for which the customer must pay. We know of one company (a leading manufacturer of convenience foods) which pays a product premium to its top vendor-rated supplier of plastic containers; the supplier's quality system is based on inspection. This arrangement serves only to reinforce the supplier's long-held belief that quality is an optional extra for which the customer must pay. Such a view ignores the financial benefits to the supplier of a continuous quality improvement process in terms of positive workforce attitudes, less waste, reduced handling, etc.

Stratified quality assurance could also be termed the 'Marks & Spencer syndrome'. Traditionally, quality assurance in the textiles industry used an inspection-based grading system which ensured that the best quality output went to a powerful customer, Marks & Spencer. But Marks & Spencer realised that if they were not actually receiving 'seconds' they were paying a high premium for products conforming to their requirement by, in effect, financing an extensive inspection system. Consequently, they encouraged their suppliers to develop a quality system based on the concept of prevention rather than detection.

A TQI strategy can provide suppliers with the organisational capacity essential for satisfying all their customers' requirements.

## Starting supplier development

Before involving suppliers in a process of supplier development it is necessary for the purchasing organisation to give attention to *internal issues* such as:

- The objectives of supplier development.
- Developing a strategy and plan to accomplish these objectives.
- Deciding which vendors to involve in the improvement plans.

They should also recognise the important role of the supplier in product and manufacturing process development and their available expertise.

An important first task is to carry out a critical review of the key aspects of the purchasing organisation's own operation which influence supplier performance. These aspects should include purchase specification, communications, training and organisational roles.

The delivery of non-conforming products from a supplier can often be attributed to an ambiguous purchasing specification. Purchasing specifications are working documents used by both customer and supplier, and must be treated as such. A good specification will define precisely the characteristics of the material to be supplied. The contents of material and product specifications have become highly standardised, and usually include such features as functional physical characteristics, reliability characteristics, methods of test and criteria for acceptance, conditions of manufacture, installation, storage and use, and so on. Sometimes such specifications are taken for granted. The purchasing department should review the accuracy and completeness of purchasing documents before they are released to suppliers. In some organisations these documents are also sent to the quality assurance department for their comments prior to transmission to the supplier, a practise we would encourage.

It is also important to recognise that the supplier is knowledgeable in his own field of operation, and should be given every opportunity to provide a design input to the preparation of the specification. Cross-functional management on the issue of new product development should involve supplier representatives on the team. This is an essential prerequisite in obtaining a supplier's continued commitment to his product after delivery to the customer. Our research indicates that suppliers are more likely to accept responsibility for defects and their associated costs if they are involved in the design of the product or formally agree the specification and drawing for the part to be produced with the customer; they can also assist in evaluating the design outcomes from their particular angle. This supplier input to the design process is a key factor in cost avoidance and reduction, and helps to reduce the product development lead time.

At an early stage in the formulation of supplier development strategy, the most effective mechanism for communication and feedback has to be identified. Typically, purchasing, quality, design, engineering/technical and production personnel all talk to suppliers, but with no single functional area accepting total responsibility for the quality, cost and delivery of the bought-out items. In some organisations, purchasing staff view assistance from the quality department in terms of dealing with suppliers as interference. The need for clear accountability is an important factor in ensuring that the channels of communication between customers and suppliers are effective. Consideration should be given by both parties to the nomination of a representative (or 'account executive') through whom all negotiations and communications for current and future business is directed. Such representatives should also be given sufficient authority to ensure that necessary actions are in fact carried out (the supplier should also nominate a person to act as the prime interface with the customer).

The supplier development process must be supported by well-trained personnel capable of helping suppliers achieve the objectives laid down. It is essential that purchasing staff understand the capabilities of their suppliers' manufacturing processes and systems, and have a good working knowledge of the philosophy, principles and techniques of TQM and quality improvement. It is also important that a customer's purchasing staff speak the same language as the supplier's quality department personnel. Our research findings indicate that only one in four companies provided quality-related skills training for their purchasing staff. Embarking on a supplier development plan with insufficient regard to the needs of the purchasing organisation's skills base is likely to result in frustration, and possibly the eventual failure of the initiative.

The increasing complexity of the task of obtaining conforming supplies at the right time and at the right price suggests that the conventional form and organisation of the purchasing management function may no longer be adequate. Traditional staff structures based on tight functional groups, e.g., purchasing, materials planning, supplier quality assurance, technical, etc., has resulted in a compartmentalised attitude to suppliers which hinders a co-makership approach to supplier development. Several companies have carried out some restructuring of their purchasing, quality and engineering departments to ensure that they have the right skills in dealing with suppliers, and that functional accountability and logistics are adequate for the task of supplier development. Good examples are the supplies management modules developed by Lucas Industries to manage the whole supplies operation from vendor selection to inventory control.[14]

**Supplier development objectives**

To assist their suppliers, some major organisations have documented the fundamental requirements for the control of quality and achievement of quality improvement, and others have even produced explanatory booklets. It is a condition of the purchase order agreement that suppliers must ensure that their product complies with these requirements. For example, the Ford Motor Company[15] World-wide Quality System Standard Q-101 – *Quality Standard Ford Motor Company* and Nissan Motor Manufacturing (UK) Ltd.,[16] *Quality Standard.* The Nissan quality policy statement to its suppliers is:

> *Nissan expects suppliers to adopt its quality philosophy, and will encourage individual and joint efforts to achieve quality improvements. Suppliers are totally responsible for their products used by Nissan and will establish quality systems accordingly. These will include continual review and development of product design, material specifications and manufacturing processes to bring about quality and productivity improvements to the shared benefit of the suppliers and Nissan. Suppliers must ensure through self-management that each product meets all the specification requirements. Suppliers are expected to operate quality control systems which meet the requirements prepared by Nissan.*[16]

Motorola Inc.[17] is following a policy of Early Supplier Involvement and as part of this have outlined their own and supplier's responsibility.

Motorola's Responsibilities

1. Match supplier with part (identify close working relations with sub-suppliers).
2. Develop contracts negotiated up-front on joint technology, developments, and commitments.
3. Provide documented assembly information, part function, application, critical design parameters, and environment.
4. Provide non-disclosure agreement.
5. Provide quality, cost, delivery, and quantity requirements.

Supplier's Responsibility

1. Provide risk assessment of selected technologies versus quality-cost-delivery.
2. Provide manufacturing input.
3. Provide advice on material selection.
4. Assist engineering in making quality, cost, and delivery trade-offs.
5. Provide itemised cost estimates for tooling, and development, based upon Motorola's inputs.

**Priorities for action**

For a company with many suppliers and bought-out items, it may take several years to introduce and develop an effective process of suppliers development. Before starting it is therefore essential to prioritise action in some way. One approach is to concentrate on new products, product and process modifications, and new vendors. Another approach involves the use of Pareto Analysis to focus priorities by ranking bought-out components and materials according to some appropriate parameter, e.g., gross annual spending. It is commonly found that some 20% of the bought-out items account for 80% of an organisation's total purchasing spending.

**Reducing the supplier base**

One outcome of the trend to co-makership is that an increasing number of major purchasing organisations are awarding more long-term contracts and contracts for the life of a part. Strategic sourcing, i.e., single or dual sourcing, is considered by many writers and practitioners to be a complementary policy to co-makership. This has, in recent years, led to reductions in the size of organisations' supplier bases. In a survey of motor industry suppliers by Dale et al,[18] only 36% had no plans to reduce their number of suppliers; the majority of suppliers had achieved at least a 10% reduction. Organisations are thinking carefully about the optimum number of suppliers they need. A typical term used to describe this is 'fewest number of most capable suppliers'. The reduction in the supplier base results in benefits such as:

- Less variation in the characteristics of the supplied product.
- An increase in the amount of time supplier quality assurance and purchasing personnel can devote to vendors.
- Improved and simplified communications with vendors.
- Less paperwork.
- Less transportation.

- Less handling and inspection activity.
- Less accounts to be maintained and reduced costs for both parties.

Nor should it be forgotten that there are competitive advantages for the supplier in being recognised as a preferred source of supply to a major purchaser. For example, obtaining the Ford Q1 Preferred Quality Award – "A supplier designated as Q1 is recognised as having achieved a level of excellence and as having in place processes and systems for continuous improvement in meeting and exceeding the customer's need and expectations."[15]

It is easier to develop a long-term relationship if the suppliers are in close proximity to the customer. Consequently, a number of customers are now reversing their international sourcing strategies to develop shorter supply lines. Closeness is also a vital element in the use of a Just-in-Time purchasing strategy.

It is also worth mentioning that it is the policy of some purchasing organisations to only take up to a certain proportion, e.g., 25%, of a supplier's output – the captive supplier issue; this sometimes results in dual sourcing even though the policy is to single source.

## The supplier development process

Having selected suitable suppliers for inclusion in their improvement plans, the next step for purchasing organisations is to get them involved and obtain their commitment. This entails making a serious attempt to communicate to suppliers what is required and, based on a set of common objectives, to reach an understanding with them.

Initially, the most practical way of setting about this task is to hold presentations to outline the new approach to suppliers, the quality system standard to be used, how suppliers' performance will be assessed and how the assessment will be communicated to them. Presentations to suppliers can be held either on the customer's premises or at an individual supplier's site. A supplier conference and/or presentation gives both parties an opportunity to air grievances and discuss problems in an open and honest manner. We have come across several examples of how major purchasing organisations communicate their improvement plans to suppliers.

- When the Ford Motor Company first re-launched Q-101 in the UK, they wrote to the chief executives of all their suppliers in an effort to secure top management commitment. The letter was followed up by presentations made by teams from the purchasing and quality departments, using a specially produced video, *Ford Cares About Quality*, to assist with the task.

- At Nissan Motor Manufacturing (UK), potential new suppliers attend a meeting where the senior management outline the Nissan philosophy and quality requirements. Those attending are encouraged to discuss frankly with Nissan's management team any areas of concern.
- IBM, Havant, take considerable care and time in ensuring that their suppliers know how their particular material, component or sub-assembly fits into IBM's final product. They hold what is termed a 'road show' where IBM representatives visit suppliers with examples of their product. The supplier's personnel are encouraged to examine the product and ask questions.

From our research, we have found that the best results are achieved when the chief executives of both the customer and the supplier are involved in face-to-face discussions.

Once a supplier's senior management team has agreed to participate in the initiative, it is necessary for the purchasing organisation to visit the supplier's factory and carry out a *formal, in-depth vendor approval survey*. The objective of the survey is to assess the supplier's suitability as a business partner. The survey is a multi-disciplinary task which in a number of cases involves the customer's purchasing, SQA and engineering personnel. The survey should cover areas such as controls, plant, technology, R&D, quality systems, attitudes, responses, tooling, planning systems and handling. Some form of check-list is generally used to structure the survey.

As part of its audit, a major purchasing organisation will assess the supplier's commitment to advanced quality planning; this is a joint exercise involving both customer and supplier, and focuses on the methods by which quality is designed and manufactured into the product. Advanced quality planning commences with a joint review of the specification and classification of product characteristics. An FMEA would also be carried out. The supplier should prepare a control plan to summarise the quality planning for significant product characteristics. This would include a description of the manufacturing operation and process flows., equipment used, control characteristics, control plans, specification limits, the use of SPC, inspection details, and corrective action methods. The supplier would then provide initial samples for evaluation, which would be supported by data on machine and process capability relating to the key characteristics identified by both parties, plus test results. Following successful evaluation of initial samples, the supplier is now in a position to start a trial production run followed by routine volume production.

Once the customer has assessed the adequacy of the supplier's policies, systems, procedures and manufacturing methods, and the supplier has been able to demonstrate the quality of his shipped product, the goods-inward inspection of suppliers can be reduced considerably – in some cases, down to the ideal situation of direct line supply. At this point, 'preferred' or 'certified supplier' status can be conferred on the supplier in recognition of his achievement.

Those companies which have made the greatest progress in supplier development operate a supplier award scheme to recognise excellent supplier performance. This is a trend which should be encouraged, but the awards must not be easy to achieve if they are to mean anything. Our research evidence indicates that suppliers use such awards in order to market themselves.

This assessment exercise is not necessarily confined to new suppliers – an increasing number of major purchasing organisations will, at regular intervals, review the adequacy of the quality system of all their suppliers. This is to assure them that conformance to the assessment awarded to the supplier quality system is being maintained. After all, most enlightened customers are looking for improvement. The frequency with which reassessments are carried out is dependent on such factors as:

- The supplier's current quality performance.
- The classification awarded to the supplier.
- The type of item being supplied.
- The volume of parts being supplied.
- The occurrence of a major change, e.g., change of management, change of facilities and process change, at the supplier.
- Request by the supplier.

A process of continuing assessment will help suppliers and purchasers to achieve quality improvement by providing a common data base.

The supplier development process does not end there; *it is an ongoing process aimed at building up an effective business relationship* – a relationship which demands a greater and quicker exchange of information between both parties. A number of major purchasing organisations are encouraging electronic data interchanges with their key vendors. The data exchange relates not only to quality, but also covers technical requirements and specifications, schedules, manufacturing programmes, lead times, inventory management, and invoicing. Suppliers are obliged to communicate any changes in materials, processes or methods that may affect the dimensional, functional, compositional or appearance characteristics of the product. Customers are obliged to provide sufficient information and assistance to aid development of their suppliers' approach to TQM (including training where necessary). In some cases this extends to joint problem-solving activities, with customer and supplier striving to improve the product and reduce its cost.

Over the longer-term it is the total cost of doing business with a supplier which is important and not the price per piece. The end result of the long-term relationship, joint problem-solving activities, and the increased level of supplier participation in the early stages of product design and development will be cost reductions to the mutual benefit of both customer and supplier.

Galt and Dale[19] report on an interesting supplier development initiative pursued by a major Japanese company located in the UK. The company was operating what they termed an Improvement Process Programme (IPP) which, at the time of their research, was operating with twelve key suppliers. The programme aimed to develop the capabilities of the supplier and bring them to a level similar of commitment and quality awareness to their counterparts, the suppliers to the mother company in its home base of Japan. It was a three year programme and included visits by supplier representatives to Japan to show them how the relationship was developing over there, and for them to assess the benefits that accrued from such a relationship. The home suppliers were also encouraged to discuss matters with their Japanese counterparts. The long-term goal of the company was to put all their suppliers through this process.

The following 'Do's and Don'ts' from Galt and Dale,[19] which they suggest should be considered by organisations interested in developing its suppliers, are a useful summary of this section.

## Do

- Look at ways of reducing the size of the supplier base. By reducing incoming material, component and sub-assembly variability, out-going product and service quality will improve.

- Ensure that, in support of the supplier development process and its various stages, your staff and those in the customer organisation use the appropriate engineering quality tools. These tools include SPC, the seven original and new quality control tools, FMEA, Fault Tree Analysis, Quality Function Deployment, and Design of Experiments. The tools also facilitate Design for Manufacturability and cost avoidance.
- Involve suppliers in new product development.
- Encourage suppliers to despatch only conforming products thereby eliminating the need to carry out duplicate testing and inspection on incoming goods.
- Award long-term contracts to key suppliers who have demonstrated a commitment to improvement in order to show tangible evidence of a long-term relationship.
- Consider implementing an assessment and rating scheme to select and measure the performance of suppliers. Poor selection will lead to increased costs, as other suppliers are sought to compensate for the deficiencies of the one chosen without due care.
- Develop procedures, objectives and strategies for communicating with the supply base.
- Treat suppliers as partners, thereby establishing trust, co-operation and dependence.
- Ensure that the staff dealing with suppliers act in a consistent and courteous manner.
- Respond positively to suppliers' requests for information.

**Don't**

- Treat suppliers as adversaries.
- Keep suppliers short of information.
- Buy goods on price alone. Ensure that other criteria such as quality and delivery performance, R&D potential, competitive manufacturing and engineering excellence are also taken into account.
- Constantly switch suppliers.
- Accept non-conforming goods.
- Talk quality, but act production schedule and price per piece.
- Forget that the initial samples procedure is a key factor in receiving conforming supplies.

## Summary

Suppliers have become an essential part of a quality improvement strategy aimed at improving the competitiveness of a buying firm. It appears that there are two major reasons for this – namely, greater global specialisation, and changes in the nature of competition itself. Effective supplier development requires purchasing organisations to treat suppliers as long-term business partners. This necessitates a fundamental shift from the traditional adversarial buyer-supplier relationship to one of co-makership, together with careful selection of suppliers and a rationalisation of the supplier base. Properly implemented, co-makership activities help reduce costs, to the benefit of both parties. There is also some technology transfer in terms of quality improvement expertise.

Demanding customers are the prime motivation for long-term quality improvement. However, the degree to which a supplier can be influenced is affected by purchasing power, and the extent to which the buyer is perceived by the supplier to be serious about a process of continuous quality improvement. Looking at industry as a whole, many major customers have started a formal process of supplier improvement, but some have no clear idea of what they require from their suppliers. Unfortunately, relatively few are actively involved in joint problem-solving activities with suppliers. Before embarking on a formal process of supplier development, the purchasing organisation must review those aspects of its own operation that can adversely affect supplier performance – issues such as purchase specifications, communications, training, organisational roles, objectives, responsibilities, priorities for action and strategic sourcing.

## References

1. Smock, D., 1982, "How to Stem the Tide of Shoddy Materials," *Purchasing*, **92** (9), 51-57.

2. Feigenbaum, A. V., 1982, "Quality and Business Growth Today," *Quality Progress*, **15** (11), 22-25.

3. Debruicker, F. S., and Summe, G. L., 1985, "Make Sure Your Customers Keep Coming Back," *Harvard Business Review*, **63** (1), 92-98.

4. Roberts, B., 1986, "The Buying Side of Selling," *Journal of Business and Industrial Marketing*, **1** (1), 69-73.

5. Philips N. V., 1985, *Co-makership: Purchasing in Benelux*, Philips NV, Eindhoven, The Netherlands.

6. Dale, B. G., and Plunkett, J. J., 1992, *The Case For Costing Quality*, Department of Trade and Industry, London.

7. Laser Media, 1988, *Statistical Process Control – Interactive Video Course*, Laser Media, Sussex.

8. Masson, R. J., 1986, "User-Vendor Relationships in the Scottish Electronics Industry," *International Journal of Quality and Reliability Management*, **3** (2), 51-55.

9. Deming, W. E., 1982, *Quality, Productivity and Competitive Position*, MIT Press, Cambridge, MA, USA.

10. Ishikawa, K., 1985, *What is Total Quality Control? The Japanese Way*, Prentice-Hall, Englewood Cliffs, NJ, USA.

11. Derose, L. J., 1987, "Changing Procurement Practices," *Purchasing World*, **31** (3), 32-33.

12. Egan, J., 5 February 1986, "Benefits of Total Quality: a Vehicle Manufacturer's View," unpublished paper presented at the Society of Motor Manufacturers and Traders *Drive For Total Quality* Conference, Birmingham.

13. Schon, D., 1971, *Beyond The Stable State*, Temple Smith.

14. White, I., and Wyatt, M., 1990, "A Case Study of Lucas Industries Approach to Supplier Integration," in Dale, B. G., and Plunkett, J. J., (Eds), *Managing Quality*, Philip Allan, Hemel Hempstead.

15. Ford Motor Company, 1990, *World-wide Quality System Standard Q-101*, Ford Motor Company, Plymouth, MI, USA.

16. Nissan Motor Manufacturing (UK) Ltd., 1985, *Nissan Quality Standard*, Nissan Motor Manufacturing (UK) Ltd., Washington, Tyne and Wear.

17. Motorola Inc., 1990, *Design for Manufacturability*, Eng 123T Course Notes, Motorola Inc., Chicago.

18. Dale, B. G., Shaw, P., and Owen, M., 1990, "SPC in the Motor Industry – An Examination of Implementation and Use," *International Journal of Vehicle and Design*, **11** (2), 115-131.

19. Dale, B. G., and Galt, J., 1990, "Customer Supplier Relationships in the Motor Industry: A Vehicle Manufacturers Perspective," *Proceedings of the Institution of Mechanical Engineers*, **204** (D4), 179-186.

# Chapter 7
# Difficulties and Barriers to Quality Improvement

## Introduction

TQI is a long-term process – it can take up to ten years to put the fundamental principles, procedures and systems into place and to create the organisational culture which is conducive to continuous improvement; it also requires considerable intellectual input by the senior management team, and a clear strategic direction and framework. Therefore it is little wonder that the vast majority of organisations do encounter problems in their improvement efforts.

This chapter explores the difficulties in, and potential barriers to, starting and sustaining the process of quality improvement, and presents the case study of an organisation not committed to TQM.

## Difficulties in starting and sustaining quality improvement

From the outset, organisations must accept that quality improvement is a long and arduous journey. Unfortunately there are no short cuts and no-one has a monopoly of the best ideas. Furthermore, once started, the momentum needs to be maintained, otherwise the gains made may be lost – our experience indicates that holding the gains is not an easy task. Even the most successful quality improvement processes have periods when little headway is made, and during these times senior management need to work hard at renewing their people's motivation and keeping interest alive.

So what are the common difficulties encountered by organisations?

Newall and Dale[1] make the point that, whilst there is considerable material on the need for TQM and its benefits, there is a lack of published information on the problems likely to be encountered in the introduction and development of a quality improvement process, and on measuring the progress made. Their research identified the following problems:

Important!

- Senior, middle and first-line management are unsure of the differences between TQM, and quality improvement, and often confuse the terms.
- Companies who purchase off-the-shelf TQM packages subsequently find that the 'package' is not ideally suited to their needs. It is then either dispensed with or extensively modified.
- A failure to plan the introduction and future development of the quality improvement process effectively.
- The stages through which an organisation's quality improvement process must pass before it is established are not identified (the same is true for expected achievements). Consequently, they are unable to monitor and assess the effectiveness and progress of the process at the various stages of its development.
- There is a lack of effective measurement criteria.
- The need for quality improvement is often questioned by functions such as R&D, Finance or Personnel, all of whom feel that it is a threat to their perceived importance and autonomy.
- There is a lack of management commitment, particularly from the ranks of middle management. The extent of this is often masked by the inadequate drive and motivation displayed by some senior managers. Any perceived lack of commitment to quality improvement from senior and middle management levels is likely to have a detrimental effect on the behaviour and attitude towards quality improvement of employees lower down the organisational hierarchy.
- Some employees have difficulty adjusting to the new and more sophisticated measurement techniques that are used to facilitate quality improvement. This has resulted in measurements being taken incorrectly and data being analysed and presented in a poor fashion. Some companies are finding it increasingly difficult to get a workforce, whose average age is rising due to the demographic shift, to accept many of the new and, to them, alien measurement concepts which are required as part of the quality improvement process.

Further data on the difficulties faced by organisations in their quality improvement efforts is provided by two questionnaire surveys carried out by Dale.[2] A total of 76% of the respondents reported that their organisations did experience some difficulty in starting the process of quality improvement. They were asked to indicate the main inhibitors (Figure 7.1). The three main difficulties appear to be:

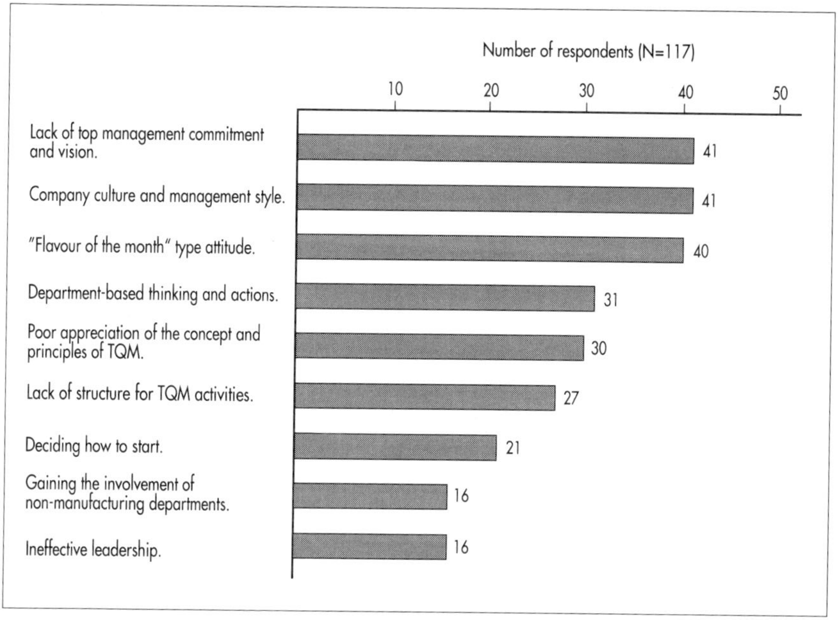


**Figure 7.1** Difficulties experienced in introducing TQM

1. Lack of top management commitment and vision (27%).
2. Company culture and management style (27%).
3. Flavour of the month type attitude (26%). For example, getting the commitment to change, people's behaviour and attitudes which are not conducive to TQM, converting the cynics and sceptics, indifference, not taking the concept seriously – people believing that quality improvement does not apply to them and if they do not respond to the TQM initiative it will eventually go away.

Some 63% of respondents also said that their organisations experienced some difficulties in sustaining TQM (Figure 7.2). The three main difficulties were:

1. Time pressures, work loads and lack of resources (37%). For example, release of people for teamwork and improvement activities, the lack of time which key decision-makers spent on TQM activities, and the chief executive officer not being prepared to devote the time to develop his personal understanding of the subject of TQM and commit the requisite resources.

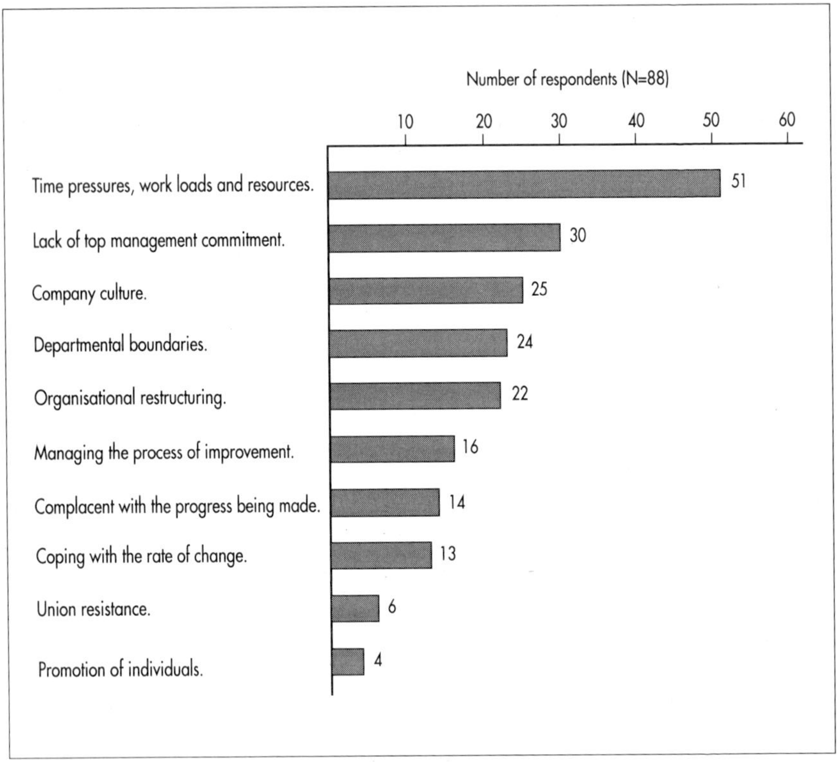


**Figure 7.2** Difficulties experienced in sustaining TQM

2. Lack of top management commitment (22%).
3. Company culture (18%).

There was no statistical evidence to indicate with any degree of confidence the existence of a relationship between the approach taken to TQM and the difficulties encountered in starting and sustaining it.

On the evidence of these two surveys and our accumulated experience there is good reason to believe that company culture, lack of senior management commitment and ineffective leadership are serious inhibitors to the introduction and development of TQM. In many organisations the board of directors and senior management team may be committed to TQM, but their words are not translated into effective action and leadership.

## Barriers to improvement

It is clear then that there are several potential barriers to the quality improvement process, barriers which, if ignored, are capable of restricting long-term benefits. Our research indicates that the main barriers can be categorised as:

- The nature of management leadership within the organisation.
- Fear of change.
- Lack of the necessary skills to facilitate improvement.
- Lack of information to support the improvement process.

These barriers are now discussed.

### Management style

In Chapter 5 we discussed how the leadership style of the chief executive shapes the culture environment of an organisation in a positive or negative direction, either through his personal actions or by neglect. At worst, unthinking, seemingly uncaring personal actions may destroy the credibility of quality improvement activities. Saying one thing and doing the opposite is a prime example of this kind of transactional leadership. This was so in the case of one chief executive, a self-declared advocate of TQM, who arbitrarily instructed that a batch of non-conforming product which was rejected by the quality assurance department be despatched to the customer on the grounds of a 'commercial decision'. By his action he was demonstrating his lack of commitment to TQM. It will come as no surprise to the reader that the despatched goods were promptly rejected by the customer's goods-receiving function. This adversely affected the chief executive's credibility in the eyes of a large part of the workforce and tended to undermine morale, including that of middle and first-line managers.

The *leadership style of many chief executives is transactional* (reacting to events) rather than transforming (shaping future events). Taking their cue from the chief executive, management teams become engaged in applying personal control, or fire-fighting, rather than planning and taking action to anticipate future events. Managers who react to events as they happen usually have poorly defined objectives, and often see the discovery of problems as an opportunity to shine as dynamic, macho trouble-shooters. The same is also true at supervisory level. Research by Trought and Willey[3] presents evidence that British managers revel in crisis management, and perceive this to be their most important task. However, personal control often deteriorates into fire-fighting, with managers spending most of their time engaged in damage-limitation exercises. Certainly there are managers in companies we have studied who attach great kudos to crisis management. In one particular case, the whole managerial ethos is management-by-crisis, which echoes the chief executive's own leadership style. This company prides itself on being 'lean and mean'. This style of leadership does not readily lend itself to the kind of planned organisational change required for effective long-term improvement.

A 'once only' input to quality improvement typifies transactional leadership. This is where the chief executive declares his commitment to product and service quality, and then promptly side-steps the issue by delegating leadership of the quality improvement process to the quality assurance manager – the 'do it but don't involve me, but keep me informed' attitude. Not only is this action an explicit denial of top management's ownership of quality improvement, but also an implicit confirmation of the common misconception that quality management is a 'bolt-on' technical activity. It is our view that although quality managers play a key role in a process of quality improvement in terms of co-ordinating, advising, training, etc., they should never be made personally responsible for managing the process. Where this is the case, it is almost certain that senior management adhere to the old attitude that it is the quality department's responsibility to safeguard product quality. The experience of companies who have been involved with quality improvement for some years is that the improvement aspect of quality needs to be separated from the day-to-day control activity, and that the improvement process should be managed by a multi-disciplinary team headed by a member of the senior management team, preferably the chief executive officer.

The findings from a questionnaire survey of automotive suppliers[4] revealed that 82% of the companies surveyed had a formal quality improvement process – sanctioned at board level in 75% of the cases. However, the chief executive was involved in the actual introduction, steering and management of the improvement process in only 25% of the companies surveyed. When the chief executive leads the quality improvement process and there is a multi-disciplinary quality improvement steering committee or council in place, two significant trends were noticeable:

- First, greater usage of what might be termed 'planning type' quality management techniques, i.e., product planning, FMEA,

formal reliability improvement programmes, project planning, vendor rating, review of contracts for special customer requirements, quality performance reporting at board level and to other management meetings, quality costing, quality circles and other improvement group activities.
- Secondly, formal quality skills training was carried out for a wider range of personnel, namely middle management, engineering staff, purchasing staff, maintenance staff, marketing staff, production control staff, and first-line supervisors.

The departments involved in the respective quality improvement processes and the status of their representatives are given in Figure 7.3.[4] It can be seen that such a process involves (in rank order) personnel from the Quality, Production, Technical/Engineering, Marketing/Sales, Purchasing, Design, Finance and Personnel departments. This suggests that some suppliers are making efforts to ensure that all departments and their personnel understand how their performance and output can affect their customers (which might be another department) in terms of satisfaction and perception of quality. In general, departmental representatives are usually managers rather than directors.

### Fear of change

Change is disruptive and disturbing. Fear of change is usually based on ignorance of the objectives and benefits of the proposed change and/or on some misconception of what is entailed. In some companies, the workforce itself, in the form of trade union representation, can be a prime barrier to change. However, much depends on the relationship which exists between management and unions. If mutual trust is missing and the union perceives change as not being in the best interest of its members, it will, not surprisingly, attempt to block the proposed change. Naturally, people are reluctant to make changes which they believe will affect their income, role, responsibilities, personal esteem and security of employment.

For example, at one company we studied, there was trade union resistance to the introduction of SPC. Their main fears included loss of earnings and jobs, the formality of the new system (with regard to personal accountability and reporting procedures), the need for increased numeracy and literacy skills, and a general worry about the unknown. The board of directors made a formal policy statement to the effect that no one would lose their job, or money, as a result

| Department/Status | | Number of respondents (N=246) |
|---|---|---|
| Quality – | Director | 11 |
| | Manager | 198 |
| | Supervisor | 17 |
| | Staff | 3 |
| | | 229 |
| Production – | Director | 43 |
| | Manager | 154 |
| | Supervisor | 13 |
| | Staff | 4 |
| | | 214 |
| Technical/Engineering – | Director | 38 |
| | Manager | 117 |
| | Supervisor | 16 |
| | Staff | 3 |
| | | 174 |
| Marketing/Sales – | Director | 34 |
| | Manager | 71 |
| | Supervisor | 2 |
| | Staff | 6 |
| | | 113 |
| Purchasing – | Director | 14 |
| | Manager | 94 |
| | Supervisor | 1 |
| | Staff | 4 |
| | | 113 |
| Design – | Director | 7 |
| | Manager | 56 |
| | Supervisor | 10 |
| | Staff | 8 |
| | | 81 |
| Finance – | Director | 18 |
| | Manager | 36 |
| | Supervisor | 1 |
| | | 55 |
| Personnel – | Director | 5 |
| | Manager | 36 |
| | Supervisor | 1 |
| | Staff | 1 |
| | | 43 |

**…onnel receiving quality skills training**

the introduction of SPC, but neither would the company pay operators for measuring and charting statistical data. It was stressed that the introduction of SPC would probably increase earnings over the long-term due to the reduction of variation and non-conforming product, thereby increasing their security of employment. The unions refused to accept this and staged a walk-out on the day that SPC was to be launched. Only a strong stand by senior management enabled SPC to be introduced.

Preston[5] argues that when demand is rising for a company's products is a better time to win support for change than when demand is stagnant or falling. The change, in terms of the introduction of SPC, at this company took place in response to a customer's contractual requirement, and the demand for their product was rising. However, senior management failed to effectively communicate this message to the workforce. If the company had involved the customer in the communication process, i.e., as proof of the need for change and the benefits of doing so, and given more thought to how to sell the benefits of the change in its SPC awareness briefings, the fear and resistance to change could have been greatly reduced.

Such fears are not confined to the shopfloor – *senior management are the people with the most power to resist change*; the degrees of freedom for change are much higher at the top of the organisation. Managerial fear of change has two root causes:

- First, change is perceived as a threat, an erosion of their authority. We have found several instances of what might be termed 'managerial intransigence' acting as a road-block to change. In these companies, quality improvement or lack of it took the shape of a guerrilla war, in which the champions of change were constantly attempting to find ways of by-passing the road-blocks. For example, in two of the cases reported in Chapter 2, top management did nothing to discourage middle management intransigence – who believed that they would lose too much output as a result of the quality improvement process, and that delivery dates would suffer. This is symptomatic of an organisation in which genuine productivity improvement is not understood, and in which the need for change has not been internalised by the majority of its people. It was noticeable that in both cases the chief executive did not act as a change agent.
- Second, there is a marked reluctance on the part of managers to examine what they do and admit (even implicitly) that their

> actions may have contributed to non-conformances – it is always someone else's fault. This attitude is most likely the result of a feeling of helplessness on the part of the individual to make a positive contribution, rather than the natural outcome of Western egocentric society.

Managers and supervisors may not be prepared for their role in the quality improvement process. The organisation's traditional environment and the individual's own experience may have bred, over the years, a directive, dogmatic and non-participative management style. However, the quality improvement process which involves company-wide employee development and participation, and teamwork in a variety of forms, requires that managers acquire the skills of listening, encouraging participation, and supporting new ideas. This participative behaviour is not really compatible with the normal leadership style of many managers and supervisors, and the result is often a severe credibility and self-image problem. Much of the effectiveness of the improvement process will be lost if middle and line managers and supervisors are not capable of functioning in their newly defined roles.

**Inadequate improvement skills**

A planned programme of training and education in quality-related skills for all employees, including members of the senior management team, is an essential prerequisite for successful quality improvement. Starting a quality improvement process with insufficient regard for the needs of the organisation's skills-base is likely to result in frustration and eventual failure because of the gap between awareness and capability. Skills of quality improvement such as the seven original quality control methods also give a company's employees a common vocabulary and help to promote an understanding of the means to facilitate improvement. It is surprising how many organisations establish quality improvement teams without ensuring that team members have adequate problem-solving skills. Employees must be taught these skills both as individuals and team members. Moreover, this type of training should be open-ended and continuous.

Based on the evidence of the findings of the survey of automotive suppliers[4] mentioned earlier in this chapter, we believe that many companies suffer from myopia when selecting personnel to receive quality-related skills training. Employees from functions such as Personnel, Finance, Marketing, and Production Control, as well as the Directorate, had not received such training in the majority of companies surveyed, (Figure 7.4.).

Our research evidence, as indicated in Chapter 4, is that many senior managers are not prepared to devote time to developing their understanding of TQM and quality improvement, and many indeed go to considerable lengths to avoid it. For example, in a recent one-day TQM awareness seminar organised by UMIST for the senior management team of a 'blue chip' company, the chief executive had instructed his secretary to interrupt the seminar some 90 minutes after it commenced to summon him to attend to a fake urgent matter which would keep him occupied for the remainder of the day. In this way, he would have shown token support for developing his understanding of TQM without being committed for the whole of the day. However, having been exposed to the first 90 minutes he was enthused by what he had heard and was an active participant throughout the seminar.

The role of the quality assurance department will become much broader as a result of increasing the board of directors' appreciation and understanding of TQM and quality improvement. We found that in those companies where the directors have received formal quality

| Position | Number of respondents (N=300) |
|---|---|
| Quality Staff | 206 |
| Shopfloor Employees | 163 |
| Supervisors | 156 |
| Engineering Staff | 113 |
| Middle Management | 112 |
| Production Control Staff | 83 |
| Purchasing Staff | 70 |
| Directors | 65 |
| Marketing Staff | 63 |
| Maintenance Staff | 47 |
| Financial Staff | 34 |
| Personnel Staff | 32 |

**Figure 7.4** Personnel receiving quality skills training

skills training, personnel from the quality assurance department are more likely to be involved in the following activities.

- Market research.
- Purchasing.
- New product design.
- Production planning.
- Industrial engineering.
- Value analysis.
- Training.
- Assisting suppliers.
- Dealing with customers.

Limiting quality skills training to the apparently traditional participants in quality improvement, i.e., quality department personnel, first-line supervisors, shopfloor employees, will be ultimately self defeating. One lesson learned from those companies which have achieved lasting success in their quality improvement efforts is that the weakest point (and one where the greatest potential gains are to be made) is where inter-departmental and inter-process boundaries cross. Not surprisingly, this weakness is exaggerated in instances where one or other, or even both, functions involved are not in possession of the relevant improvement skills.

In general, little thought is given to designing quality-skills training programmes for employees with less tangible measurable outputs than, for example, shopfloor operatives. This is hardly surprising, for while the interactive service-chain concept of quality management, (i.e., for each person in the organisation, the next person or department to use their output becomes the customer), is much talked about in terms of the process owner being responsible for ensuring that the customers' requirements are clearly understood, it has so far made relatively little impact on the practice of European industry; few companies consider process inputs and outputs and process owner needs in their non-manufacturing activities. Moreover, the methodology of process modelling is not yet widely understood.

*The skill needs of the quality improvement leaders are often taken for granted.* In the majority of cases, the task of co-ordinating and advising on quality improvement is delegated to quality assurance managers, and whilst they might be in possession of the necessary techniques and related skills of quality management, they may well lack the vital communications, selling, teaching and project-management skills needed by an effective improvement leader. Such skills may not necessarily be available in a convenient training package and they need to be developed.

As mentioned earlier, the ideal solution is that the task of managing the quality improvement process be given to a multi-disciplinary team, chaired by the chief executive officer. However, there is a danger that unless its members are genuinely committed to employee participation and skills development, the appointment of such a team will merely create what the rest of the company regard as an elite, so preventing the transfer of ownership across all levels and functions of the organisation; the improvement process remains 'theirs' not 'ours'.

There is often a tendency to launch too many quality initiatives, resulting in departments and quality improvement teams having too many live projects on the go at any one time. We know of one manager in a telecommunications company who is currently engaged in 42 different improvement projects. This tends to deplete resources and could be avoided by better project management.

**Poor information**

The effectiveness of the quality improvement process is dependent on the availability of relevant, reliable and objective information. This statement is perhaps self-evident: however, as with most axioms, it is usually taken for granted. The kind of information required includes:

- Data on key product characteristics and features.
- Process parameters.
- Process variation.
- Process objectives.
- Process inputs and outputs.
- Quality costs.
- Internal and external performance measures.
- Customer requirements, etc.

This data is often unavailable.

The organisation is then faced with two problems:

- First, it has no datum line from which to measure improvement.
- Second, aspects of the business which are not measured are difficult to control.

In many instances, senior management appear to be unaware of the seriousness of the situation, usually because the lack of real information is camouflaged by a flow of extraneous data. In several companies we have studied there was a considerable amount of quality-related data in the form of scrap reports, warranty reports, engineering change documents, vendor rejection notes, and so on; but there was no objective information on the causes of process variation and vendor non-conformance, the cost of non-conformance, direct measures of customer satisfaction, etc. In all these companies, formalised quality improvement activities did not take place either because senior management believed that they had the situation under control or because they were ineffective as time and resources were wasted chasing phantom problems. Clearly, poor information is harmful to the quality improvement process.

Price[6] defines three rules for systematic quality improvement:

1. No measurement without recording.
2. No recording without analysis.
3. No analysis without action.

Failure to obey these three simple rules can mean that information essential to the quality improvement process, which is often available within the organisation, is lost. This is an area to which management need to pay serious attention. The following are typical scenarios:

- The information is easily obtainable but not recorded, or it may be in the possession of people perceived as remote from the mainstream of quality management activity and thus overlooked.
- The data is recorded but is not used; for example, no attempt is made to analyse cost information in order to attribute quality costs to department, business unit, defect type, product, supplier, customer, etc.
- A great deal of time and effort is spent in developing control charts to listen to the voice of the process, but once the data has been collected the company does not use the data to improve the process.

There is little doubt that the flow of quality-related information within an organisation is often disjointed. The subject of quality information systems and database management is an area in which further research needs to be carried out.

## 'Brewco': a story of non-commitment

### Background

The behaviour and attitudes of managers in organisations committed to TQM is well documented.[7-10] *What then is the state of quality awareness, behaviour and attitudes of management in an organisation where the normal agents of change are either missing or diluted?* In other words, an organisation whose management have no apparent incentive to direct their efforts to improve product and service quality. We have carried out a study in such an organisation; there now follows the story of 'Brewco'.

Brewco is a brewery located in the UK. In common with most of its competitors during the last two decades the company became more of a general leisure business, with interests in hotels and catering, although beer production still accounted for approximately 80% of the company's annual turnover.

The brewery's quality management organisation was divided into two functions: internal quality control and external quality control. The former was headed by the technical manager and included the laboratory. In addition to a laboratory manager, there was also an internal quality control manager whose prime occupation was the recording of product attribute data, e.g., flavour, colour, clarity, etc. There was a separate Quality Control department responsible for external quality assurance, that is, the quality of the beer after it has left the brewery. This department, headed by the quality manager and assisted by the quality assurance manager, provided licensees with technical assistance with regard to product quality and managed the trade credit scheme for returned beer. The roles of these five individual managers were not precisely defined, and their responsibilities often appeared to be unclear. Furthermore, communication between the internal and external quality management functions tended to be informal and *ad hoc.*

### Senior management attitudes and perceptions

From interviews carried out with a cross-section of the company's managers, it soon became apparent that there was a wide gulf in understanding between the brewery, i.e., production, on one side, and the sales divisions on the other. The contrasting views of two senior managers (the tenanted sales manager and the brewing manager) from either side of this divide are illuminating:

The tenanted sales manager said that the greatest problem affecting sales was product variability. He claimed the beer lacked consistency and this was reflected in the consumers' attitudes to the product, which was often negative. He felt that senior management in the brewery did not appreciate the problem, and as a result licensees were often encouraged to sell beer even though customers had complained. Discussions revealed that this senior sales executive not only had practically no product knowledge, but absolutely no appreciation of what actually went on in the brewery. Furthermore, the only regular communication he had with the brewery, situated approximately 800 metres from his office, was by memoranda.

The brewing manager stressed the inherent variability of the raw materials, claiming that achieving consistent product uniformity was not easy with what is essentially an agricultural product. He stated that independent test results (by the Brewing Industry Research Foundation) had shown Brewco's products to be fairly consistent. He said the problem was that other departments were often too critical of the brewery's output. For example, the packaging department complained of sediment at the bottom of casks – he maintained that such perfection was not possible. The sales divisions had unrealistic requirements and always sided with customers against the brewery. Sales representatives often created product quality scares in the trade by latching onto a product complaint in one pub and then going round asking other customers if they had the same complaint. This kind of action often planted the idea in licensees' heads that something was wrong with that particular batch of beer, the result being a rush of complaints and requests for replacement beer. He clearly believed sales personnel to be irresponsible.

If there was ever a clear illustration of the need for Deming's Ninth Point[11] – 'break down barriers between departments' – it was here.

Within Brewco there appeared to be a fundamental paradox: how to reconcile a naturally variable agricultural product with the needs of a sophisticated market demanding a consistently uniform product. Other interviews supported our belief that this apparent paradox was not only deeply embedded in the minds of most of the company's senior management, but also acted as a barrier to change. It seemed as if in spite of the existence of a state-of-the-art computer controlled plant, a collective belief that brewing is a 'black art' still prevailed. Moreover, customer satisfaction was generally perceived as a subjective exercise at best. This attitude was not helped by the sales divisions' apparent inability to provide positive input for the brewery with regard to either objective measurement of customer satisfaction or product performance after delivery. The problem was exacerbated by the following factors:

- A general lack of perception throughout the company of how all operations affected product quality. Product non-conformance was generally perceived to be the result of what went on in the brewery, whereas in reality the brewery's operations were just one of three elements of product quality: the other two being raw materials and the distribution process. All three elements were recognised and commented on separately, but very few managers interviewed linked all three together. In other words, most of the company's senior management did not appear to appreciate the scope of their potential quality problem and subsequent improvement.
- No mechanism existed with which the scale of the potential quality problem could be measured. Whilst quality-related data such as batch sample results, flavour profile tests, and returned beer reports were recorded, these were neither analysed nor acted upon.
- The company had neither a documented and objective quality system along the lines of the ISO 9000 series of standards, nor a well-defined communications infrastructure. Senior managers were prepared to ignore quality parameters and specificational requirements when it suited them. The technical manager complained of a lack of discipline amongst his colleagues, and stated that most of them had dual standards and were prepared to despatch beer which was clearly out of specification if, in their opinion, it was saleable.
- The management structure of the company was such that individual departmental managers had considerable discretionary power and appeared to manage their departments as though they were independent sovereign states. Naturally, this environment was hardly conducive to company-wide co-operation or joint decision-making. Furthermore, quality-related activities were seen as a departmental on-cost to be minimised as far as possible by managers preoccupied with their own cost centres rather than with overall company profitability.
- The production director appeared neither to appreciate the benefits of effective quality management nor be prepared to admit the possibility of anything being wrong in the brewery.

- Lack of proof of the need for quality improvement within the brewery as the direct effects of market pressure and customer purchasing-power was absent. Large profits, a massive positive cash flow, and the tied-house system cushioned the brewery from the changing realities of the market.

**Attitude survey**

A questionnaire survey was carried out to gain some insight into the awareness and attitudes to quality of the company's management and sales teams in order to put both the perceived attitudes of the senior managers interviewed and our own empirical observations into perspective.

The survey revealed a positive attitude to product quality. All the respondents not only thought that quality should be at the top of the company's corporate objectives, but also that it was important to them personally to be associated with a company which had a good reputation for quality.

The main findings of the survey can be grouped around the following themes:

1. *Quality as a corporate priority.*
   The majority of respondents (78%) believed that quality lay at or near the top of the Brewco's corporate objectives. Only five respondents were dissatisfied with the quality of the company's products. The majority of respondents (75%) were fairly satisfied with the company's efforts to improve quality. The company was not perceived as placing any special emphasis on meeting production volume targets or keeping down costs at the expense of product quality. But only four respondents believed that employees were given sufficient incentive to improve quality within the company. The consensus of opinion appeared to be that the company's quality management efforts were satisfactory, but it could do better.

2. *Personal responsibility for quality.*
   The majority of respondents acknowledged that they had a personal responsibility for product quality and were involved in quality related decision-making. Over 75% of respondents thought that they had at least 50% of the responsibility for quality in their job, although, paradoxically, a significant number of respondents (57%) believed that at least 75% of the responsibility for product quality belonged to Quality Control – which strongly indicated a low aware-

ness and poor definition of quality management responsibilities. Most respondents (83%) stated that they relied on other people doing their job correctly to enable them to do their own job correctly, thus acknowledging the cross-functionality of quality management, i.e., the next process owner is the customer.

3. *Poor information and feedback.*
   Earlier in this chapter we cited poor information as a barrier to improvement. Our sample of Brewco personnel commented on the problem of poor information and feedback. The majority of the respondents (52%) claimed that they received inadequate feedback about quality in their work. Typical comments included:

   > *As each operation from brewhouse to customer is departmentalised it is difficult to assess the quality of our own work. The Quality Control department should provide an information flow from customers back through each production department.*

   > *More information should be made available on trends and performance with regard to quality.*

   > *Closer contact is required between the Production and Sales departments in order to ensure continuity of feedback and subsequent action to remedy any problems.*

4. *Quality-related training.*
   Only half of the respondents believed that the company provided them with enough aids and sufficient skills to enable them to do their job right first time. As many as 80% of respondents thought that quality-related training would help them do their job better. But, interestingly, when analysed by department there was significant variation in response. Some 90% of respondents from Production stated they had sufficient aids to do their job, and only 40% of respondents from this department thought that quality-related training would help them to do their job better. Statistically significant tests indicated that personnel from departments not directly involved in the production of bulk beer, especially those perceived as isolated from the mainstream of manufacturing activities, e.g., Sales, were more likely to feel they would benefit from quality-related training. Several comments were made by respondents to this effect, such as:

*More knowledge of the systems used should be made available to sales staff on a regular basis.*

5. *Departmental alienation.*
A number of respondents stressed the lack of company-wide co-ordination and cross-functional communication.

One typical comment was:

*A greater effort should be made to bring all sectors of our business into the quality family. One suggestion could be to hold regular inter-departmental quality workshops. I am certain that these would highlight problems and give people a greater awareness of the need to work together.*

Some respondents even gave the impression of feeling alienated from their colleagues in other departments. This was exacerbated by the wide gulf in perceptions between the Production and Sales functions, as reflected in some of the comments made by respondents, reaffirming the attitudes of their senior managers. For example, typical comments made by respondents from the Production department included:

*Greater backing required from persons at the consumer end of the business.*

*There should be a more positive approach to quality from our representatives in trade, and especially in cellar and bar hygiene, which ultimately has a direct effect on quality.*

By contrast, the following comments were made by respondents from the sales divisions:

*Sometimes the product arrives at an outlet under par and in most cases the licensee is encouraged to sell it anyway. This situation should never arise.*

*Quality should be ascertained before the product is distributed, not afterwards. The brewery should respond to constructive criticism and not pretend matters are acceptable.*

*More internal quality control; stop the problem at source.*

In summary, respondents perceived that they had personal responsibility for quality but an unclear focus on their role in quality improvement activities; a focus that could be sharpened by better information and communication, training and cross-functional process management. However, progress in this respect would be inhibited as long as inter-departmental (almost tribal) barriers and misconceptions remained. The trouble was that two essential ingredients were missing: *top management vision and a collective sense of purpose.*

**Subsequent developments**

Follow-up visits to Brewco revealed that the complacency which was evident among senior management had started to ebb away. There were several reasons for this, chief of which was a shift in the balance of power.

Traditionally, the brewery's role as a production function had been pre-eminent, the head brewer/production director was the single most important figure in the company, and the sales function was considered to have a minor role in the scheme of things, e.g., tenanted sales representatives were seen as little more than rent collectors. This view still appeared to exist in many parts of the brewery even though the business environment had changed; the market had fragmented and the days of long production runs were over. This resulted in the company having to become more sophisticated in its marketing and new business development activities. In fact, the brewery's contribution to the Brewco's overall profit has steadily declined in recent years. The net result is that the sales divisions are now pre-eminent. Indeed more than one main board director expressed a desire to see the brewery isolated by hiving-off all the pubs into a separate retail company.

So it was against this background of a gradual shift of power from the brewery to sales that sales were starting to press home their advantage. Criticism of the brewery by sales personnel had hitherto been implicit and *ad hoc*, but the sales divisions were by now openly accusing the brewery of incompetence.

Possibly the most serious criticism of the brewery came in a memorandum written by the managing director of the retail management division. This read as follows:

> *Attached is a beer report done on some tank beer tasted on . . . It was definitely not at its best and must have caused customer dissatisfaction.*

*As you will see from the notes on the analysis, this is considered to be okay in the brewery. I think this shows how apathetic we have become in that we are satisfied with a very wide specification rather than having the objective of a 'target' quality. Surely we must strive to achieve excellence not mediocrity.*

Senior management in the brewery were shaken out of their complacency. They became desperate for solutions to improve quality. But their actions appeared to verge on widespread panic involving heated discussions on returned beer analysis reports, ad hoc changes in process controls, and making wild guesses on the causes of process variation, whilst at the same time still officially refusing to admit that anything was wrong. Their reaction was one of fear compounded by external threat. Basically a lack of objectivity and a lack of information prevented the formulation of a coherent and concentrated improvement strategy.

A further development was the appointment from outside the company of a new sales director to take charge of the tenanted houses division. His ideas on personnel involvement, communications and training appeared to be radically different from those of his predecessor and many of his senior colleagues. In pursuit of his objective of improving the efficiency and professionalism of the division, he introduced regular monthly meetings and seminars for all of his area managers and representatives. As greater product knowledge and understanding of quality management and improvement were two of the goals selected by the sales director for his division, he asked the quality assurance manager to participate in the training sessions. This was the first indication of positive action to bridge the gulf between the brewery and sales.

Another development concerned the provision of objective information. Product faults were classified into two groups: the brewed and packaged commodity, and mechanical and human failure. This was the first attempt by the brewery to discover and quantify the fundamental causes of product non-conformance.

These three recent developments (sales pressure, inter-departmental co-operation, and a potential quality database) may have proved to be the key to quality improvement at Brewco. It is not inconceivable that changes in senior personnel would have helped speed-up the process of attitude change.

The final development. Readers will notice that we write of Brewco in the past tense. That is because Brewco is no more – it no longer brews beer. The brewery has been closed down and beer is bought from other breweries. Brewco lost the competitive race – a race its senior managers did not even recognise.

**A final comment**

Brewco is a prime example of an organisation where changes relating to quality improvement were very slow because there were no compelling perceived reasons for change, and where, as a result, managerial complacency was widespread. Not only was proof of the need for change missing, but a lack of information and objectivity about activities and processes within the organisation reinforced managerial complacency and prevented even minor improvement initiatives. Additional factors stifled the development of the quality improvement process:

- Absence of effective quality assurance and system standards.
- Inadequate product knowledge in many parts of the organisation.
- Lack of objective feedback on customer satisfaction.
- A parochial view of the production process and of the industry, in general, by most of the management team.

The findings of the attitude survey support the commonly stated view in the literature on TQI that 'no one is against quality', and thus behaviour and attitudes to quality are intrinsically positive even if awareness is poor.[12] This provides a firm basis for change if managed in the right way, although it is tempered by inter-departmental alienation and a common belief that the management of quality is the responsibility of the quality assurance function, which should act as a focal point for the improvement process. However, Brewco's quality function was clearly in no condition to take the lead, even if this role concept were valid, which it is not.

Before the closure of Brewco a number of developments, e.g., legislation on the number of public houses owned by different individual breweries to stimulate increased competition, the threat of a takeover bid, discussion about closure of the brewery, and the Single European Market, demonstrated that even complacent organisations cannot operate indefinitely as though in a vacuum, and that ultimately market pressure is the only argument which convinces senior management to commit time and resources on quality improvement activities. As previously mentioned this pressure, combined with the poor state of health of quality improvement, came too late to save Brewco.

## Summary

TQI is a long-term process which requires considerable dedication and hard work to achieve the vision. Along the journey there are many pitfalls to avoid and barriers to surmount, these include inadequate leadership, fear, inadequate quality and problem-solving skills and poor information. Top management commitment, motivation, training, data analysis, and effective communications media are on their own insufficient to bring about real and lasting improvement; in any case organisations, all too often, just pay lip-service to these activities. As we saw in the case of Brewco, whilst the key ingredient of TQI – people with a sense of pride – is usually already in place, if the key change agents, customer pressure and the chief executive officer are are missing the change process will not begin. Top management vision and a collective sense of purpose to achieve a common goal – customer satisfaction – are essential.

## References

1. Newall, D. and Dale, B. G., 1991, "The Introduction and Development of a Quality Improvement Process: A Study," *International Journal of Production Research*, **29** (9), 1747-1760.
2. Dale, B. G., 1991, "Starting on the Road to Success," *The TQM Magazine*, **3** (2), 125-128.
3. Trought, B., and Willey, P. C. T., 1986, "A Comparative Study of Lateral Work Activities," in McGoldrick, P. F. (Ed), *Advances in Manufacturing Technology*, Kogan Page, London.
4. Lascelles, D. M., and Dale, B. G., 1988, "A Study of the Quality Management Methods Employed by UK Automotive Suppliers," *Quality and Reliability Engineering International*, **4** (4), 301-309.
5. Preston, G., 1986, "Organisational Development as a Base for Total Quality Assurance," *Proceedings of the 25th Quality Assurance Conference*, Institute of Quality Assurance, 63-67.
6. Price, F., 1984, *Right First Time*, Gower Press.
7. Smith, S., 1987, "How to Quantify Quality," *Management Today*, 86-88.
8. Van Ham, K., and Williams, R., 1986, "The Quest for Quality at Philips," *Long Range Planning*, **19** (6), 25-30.
9. Barker, R. L., 1988, "Total Quality Management – The Only Way Forward," *Proceedings of the First International Conference on Total Quality Management*, IFS Ltd., Bedford, UK, 49-60.

10. Anderson, D. N., 1989, "Quality; A Positive Business Strategy," *Proceedings of the Second International Conference on Total Quality Management*, IFS Ltd, Bedford, UK, 201-211.
11. Deming, W. E., 1982, *Quality, Productivity and Competitive Position*, MIT Press, Cambridge, MA, USA.
12. Crosby, P. B., 1979, *Quality Is Free*, McGraw-Hill, New York.

# Chapter 8
# Quality Awareness, Communication and Challenges

## Introduction

The government in the UK made its commitment to improving quality performance in British industry crystal clear when, in April 1983, it launched the National Quality Campaign. The purpose of the campaign was to:

- Enhance the UK's reputation for quality in world markets.
- Make management and employees more aware of the strategic importance of product and service quality.
- Encourage adoption of the appropriate systems and techniques.

The campaign consisted of a number of initiatives including:

- Quality management awareness presentations.
- Preparation of booklets and videos.
- Financial assistance to meet the requirements of the BS 5750 quality system series, etc.

This chapter reports the main findings of a study into the effectiveness of the UK's National Quality Campaign. It also provides a summary of the UK Department of Trade and Industry's (DTI) subsequent initiative 'Managing into the 90's' which subsumed the national quality campaign. The chapter then examines the concept of 'technology transfer' in relation to quality management and explores the European and American plans for, and experiences relating to, improving TQM expertise and business performance.

## UK National Quality Campaign

The Manchester School of Management, UMIST, was commissioned by the Quality, Design and Education Division of the DTI to study the impact of the National Quality Campaign on British industry.

The DTI was particularly interested in the effectiveness of the campaign, and the attitudes and opinions of people in industry on the subject of quality management. The main findings of this study,[1] are as follows:

1. *The Campaign was relatively successful in reaching senior management.*
   Early in the National Quality Campaign an extensive range of publicity brochures and videos were commissioned and produced by the DTI. These were advertised in a government publication entitled *Getting to Grips with Quality* (frequently shortened to 'Grips').[2] The first issue of 'Grips', published in 1985, contained some 32 different topic brochures and nine different videos which were available free of charge from the DTI (the videos were available on loan).
   
   Over 30% of all the requests for brochures and videos received from industry in that first year of publication were from board members (Figure 8.1). Furthermore, in a survey of chief executive officers described in Chapter 5, more than half of the respondents stated that their perception of the importance of product and service quality in international competitiveness had changed during the period 1984-88. Increased general awareness of quality via the

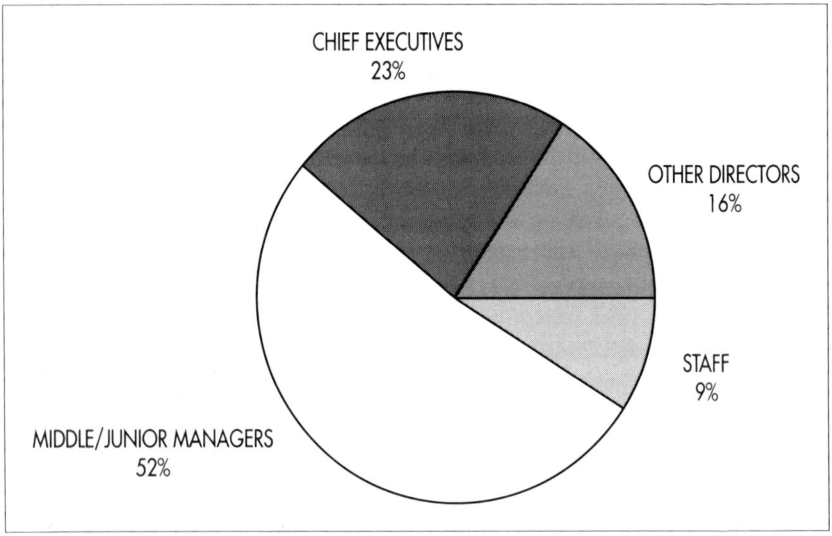


**Figure 8.1** Responses to 'Grips' by seniority

National Quality Campaign, customer pressure, and competition were the most frequently cited reasons for the change.

2. *There was a strong correlation between choice of publicity material and respondent position.*
Selection of campaign brochures appeared to be dependent on three factors:

 - Respondent's position in the company hierarchy.
 - Brochure content.
 - Respondent's perception of its relevance.

There were two major clusters of brochures reflecting the respondent's position and main interests in TQM: brochures regarding general business topics and money, and brochures covering techniques and training. Chief executive officers and directors tended to request the former cluster and middle/junior managers and functional staff the latter. For example, chief executive officers showed significant interest in certain brochures intended specifically to help them improve or maintain business performance. They also displayed significant interest in those brochures concerning financial assistance for developing their quality system to meet the requirements of the BS 5750/ISO 9000 series.[3] Managers, on the other hand, appeared more interested in brochures relating to specific quality management tools and techniques which can help them manage quality and facilitate improvement on a day-to-day basis.

3. *Campaign publicity material was used to increase in-company awareness.*
Respondents requested material for more than one reason (Figure 8.2). Nearly half did so to develop a personal understanding of TQM; over one-third did so to determine the availability of additional information on TQM and associated tools and techniques and a similar proportion wanted to find out how to obtain quality system accreditation to the requirements of the BS 5750/ISO 9000 series. However, for the majority of respondents (67%), the major motivation for obtaining material was a desire to increase quality awareness within their respective organisations. At least one-third of the respondents obtained material specifically for use at an internal seminar or TQM presentation.

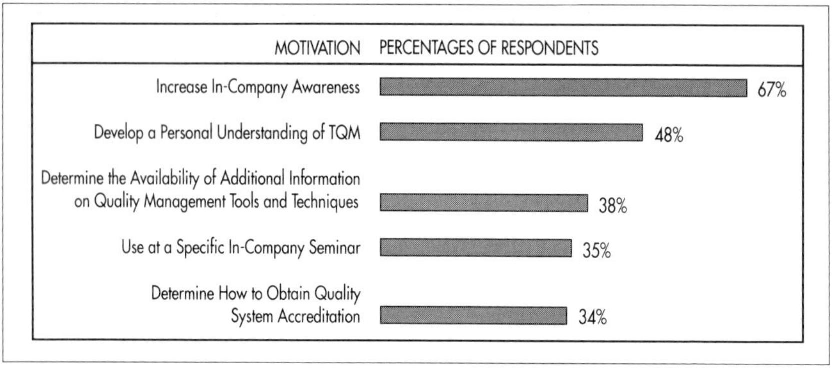


**Figure 8.2** Motivation for requesting NQC material

After receiving the material, most of the respondents (86%) read the brochures and/or viewed the videos. In around half the cases, the material was passed to colleagues and/or subsequently used in the organisation.

In a subsequent survey carried out for the DTI,[4] we found that 87% of respondents believed that the campaign had helped increase internal awareness of quality.

4. *Following the initial publicity aimed at raising awareness, there was need expressed for practical 'how-to' learning material.*
Almost half of the respondents stated that they required follow-up material to reinforce quality awareness in their companies in the form of additional information on specific quality improvement tools and techniques or industry-specific 'how to' material. Respondent needs included:

- Self-use training packs on problem-solving and SPC.
- An explanatory guide on how different quality improvement tools, techniques and systems link together.
- A practical 'how-to' guide on how to start the quality improvement process.

For example, one respondent described his company's needs as follows:

> *More practical information on how to start the quality improvement process. What are the obstacles and challenges? We have now found these out the hard way. It would have been more effective to have had a short-cut available.*

We believe that this statement articulates the sense of isolation which affects the management teams of many companies wishing to initiate the process of quality improvement.

5. *The campaign was perceived as having a beneficial impact on company performance.*
   A total of 70% of respondents believed that the campaign would benefit the overall business performance of their company.

6. *Respondents had high expectations of the government's role.*
   Respondents were given a list of interested parties who might be expected to be prominent in advocating and encouraging companies to use improved methods of quality management, and requested to rank in descending order of importance the four parties they considered most important and influential.

   As can be seen from Figure 8.3, managing directors scored by far the most ranking points, followed by the government, large purchasing organisations, quality managers, and the British Standards Institution. If grouped together, large purchasing organisations and large companies ranked in second place. This finding reinforces the hypothesis outlined in Chapter 3 that the chief executive officer and demanding customers are in combination the most powerful change agents for quality improvement; the chief executive officer must provide leadership, and direct market pressure is demonstrable proof of the need for improvement. But it is interesting to note the high ranking given to the role of government.

   The respondents had high expectations of the government's role in continuing to promote national awareness of quality management methods and disseminating related information. This is underlined by the fact that approximately 75% of the respondents received their copy of 'Grips' via the National Quality Campaign mailshot, and almost one-third originally heard of the campaign through government literature. A number of respondents requested more direct government support for the campaign in the form of, for example, tax incentives for companies on additional profits resulting from quality improvement activities, and mandatory quality requirements for industry through government legislation.

## An overview of the campaign

If the prime objective of the National Quality Campaign was to increase industry's general awareness of the importance of TQM, then

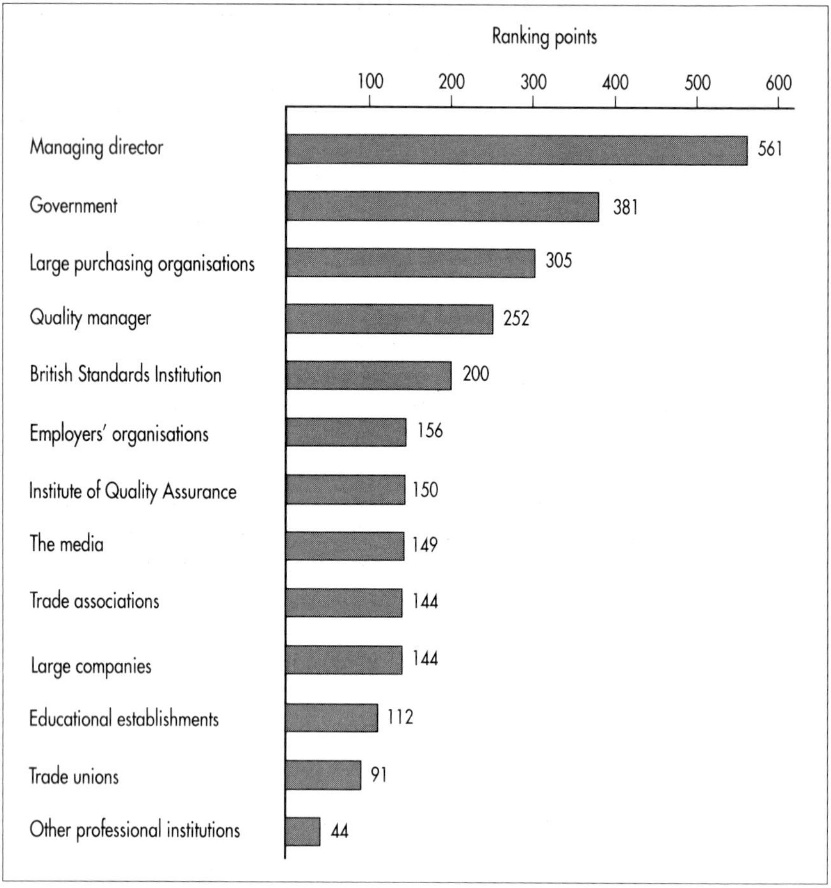


**Figure 8.3** TQM influence groups ranking

it can be considered to have been a success, not least because the government's involvement has given credibility to the message.

However, an increase in awareness of the need for quality improvement is not, on its own, enough to provoke a revolution in the way that British industry manages product and service quality. According to the DTI publication, *Quality Counts*, the overall aim of the National Quality Campaign was:

> *. . . the enhancement of the quality of British goods and services through encouraging a total approach to quality in design, production, marketing, delivery and the way a business is run.*[5]

It is no exaggeration to say that this is an ambitious goal, requiring a fundamental change in attitude on the part of much of British industry. A goal based mainly on the government's ability to persuade British industry to adopt the series of quality standard BS 5750. The latest edition of the DTI's *Register of Quality Assessed UK Companies* contains over 15,000 entries which were assessed under the British Standard Institution and similar schemes (Lloyd's Register of Quality Assessment, Department of Health, etc.) as well as those run by such purchasing organisations as British Coal, British Nuclear Fuels, British Telecom, the Ministry of Defence, etc.[6] In addition, many more companies have had their quality systems subjected to second-party assessments, often against the requirements of the BS 5750 series. Fifteen thousand companies may not appear to be a high figure, but at least it demonstrates the growth of demand for quality system assessment.

A key question is whether the government should still continue to play an active part in helping industry to introduce and develop the process of quality improvement. The government's involvement in the National Quality Campaign has been generally recognised as a crucial factor in industry's awareness of the need for improved product and service quality. Indeed, the findings of the survey of 'Grips' respondents revealed that the majority still expected the government to play a leading role.

The DTI demonstrated its continued commitment to industry-wide quality improvement when, in January 1989, a new programme called 'Managing into the 90's' was launched under the umbrella of the Enterprise Initiative.

## Managing into the 90's

'Managing into the 90's' provides a complete programme to help companies build-up their competitive performance and improve the overall effectiveness in four key business areas:

- Design.
- Quality.
- Purchasing and supply.
- Manufacturing management.

As part of the programme, the DTI is running management briefings, workshops, demonstration company visits and mobile demonstration units to provide industry with more information and advice on how to integrate these four key areas into a coherent strategy for continuous improvement. The programme focuses on the cross-functional aspects of design, quality, purchasing and supply, and manufacturing management.

'Managing into the 90's' is an important programme which recognises that an increasing proportion of British industry has reached the stage of wanting to translate enhanced quality awareness into improvement action. Upon reaching this stage, company management teams typically want to know:

- What to do.
- What techniques to apply.
- How to overcome the obstacles to progress.

Whilst the campaign gives some broad brush guidelines, there is a limit to the government's effectiveness in influencing the activities of private-sector industry. Government-sponsored activity cannot offer more than a partial and transitory assistance to the dynamic environment of industry. Companies must accept intellectual ownership of the TQM concept and devise their own quality improvement strategies and initiatives. There has been a tendency to see the National Quality Campaign purely in terms of a government economic strategy. But the DTI does not possess the kind of influence enjoyed by, for example, the Japanese Ministry of International Trade and Industry (MITI) to enable it to single-mindedly co-ordinate the competitiveness of the national economy.

The task facing those who are keen on helping European industry increase its competitiveness through TQM and the process of quality improvement has changed from propaganda to one of establishing a framework for successful implementation. In other words, the new task is that of technology transfer.

The term 'technology transfer' means the utilisation by organisations of technology skills or knowledge from external sources. Technology transfer has become an important mechanism for business development, and such transfers can have important economic and growth implications. However, many firms (not just small ones) are constrained in their use of available resources by time, lack of knowledge, lack of experience and lack of understanding concerning how new technological advances might benefit them. It could be argued that TQM is one such technological advance (even though the concept is by no means new), and that therefore the adoption by an organisation of a process of continuous quality improvement is technology transfer.

We believe that resources should be committed by countries to developing a national infrastructure for the transfer of TQM-related technology, devising mechanisms for the exploitation/sharing of scarce quality management resources and as a pan-European dimension.

## An infrastructure for technology transfer

### The need for partnership

The essential infrastructure for effective technology transfer requires a partnership among government, industry (especially in the utilisation of the purchasing power and resources of major industrial companies in the private sector), and other interested parties, e.g., educational establishments, professional institutions, etc. It is fair to say that so far in the UK there has been little progress in marrying these disparate elements into an effective infrastructure. The DTI's publication, *Caring Enterprise and Winning World Market Share*, describes the need for a new foundation to co-ordinate activities and act as the focal point of a national initiative to stimulate the 'pursuit of excellence'.[7]

The Union of Japanese Scientists and Engineers (JUSE) provides a well-founded model for an establishment to act as the cornerstone of a TQM transfer infrastructure.

### Japanese Union of Scientists and Engineers

JUSE is acknowledged as the centre for Total Quality Control (TQC) in Japan. JUSE was established in 1946, and in 1962 was consolidated under the jurisdiction of the Science and Technology Agency of the Japanese Government to cope with the advancement of society. The influence of JUSE on the development of TQC in Japanese companies through its extensive structure of TQC education and training courses is considerable. This influence and the criterion of the Deming Application Prize (company policy and planning; organisation and its management; QC education and dissemination; collection, transmission and utilisation of information on quality; analysis; standardisation; control; quality assurance; effects; and future plans) has resulted in a very consistent approach being taken to TQC.

JUSE endeavours, as its basic objective, to contribute to the industrial sector for the advancement of science and technology through its various activities,[8] including:

- Education and training – regular training courses are held in quality control, reliability, design of experiments, multivariate analysis, operations research, industrial engineering, marketing, sensory testing, product liability and other management techniques, e.g., cost reduction, production control, value engineering and finite element methods.

  They offer 27 different courses which deal with TQC and a further 9 relating to reliability. These are attended by a range of people from presidents and top management to line workers. At JUSE, between 600 and 800 people receive TQC training on any one day. Table 8.1 shows examples of the courses held. A striking feature of the data contained in the table is the duration of the training courses, the length of time some of them have been in existence and the number of participants who have attended them.
- Research and development – research on new scientific technology as applied to corporate management. Special committees are established to deal with major subjects, e.g., the seven new quality control tools, research is commissioned by government agencies and companies, and joint research with companies and universities on specific themes is undertaken.
- Symposia and conferences – these include, quality control, QC circles, reliability and maintainability, sensory inspection, multivariate analysis, the seven management tools for quality improvement, product safety, numerical analysis in fluid mechanics, quality control of software production, and quality control of service industry and ergonomics/human factors.
- Technical counselling – quality control diagnosis, instruction at the workplace and in-company education and training.
- International co-operation – information exchange, personnel exchange, sending and receiving of study missions, dispatching of lecturers and receiving trainees, and holding of international conferences on TQC.
- Extension services – the following awards and nation-wide organisational activities are established and supported by JUSE:
  - Deming Prize.
  - Japan Quality Control Medal.

| Subject | Period (days) | Frequency times and classes | Cumulative number of Participants (as of March 31st 1990) | Establish (year) |
|---|---|---|---|---|
| QUALITY CONTROL | | | | |
| QC Basic Course | 30 | 7 | 24823 | 1949 |
| QC Middle Management Course | 12 | 10 | 28632 | 1955 |
| QC Top Management Course | 5 | 5 | 4327 | 1957 |
| QC Introductory Course | 8 | 7 | 23728 | 1957 |
| QC Executive Course | 5 | 6 | 7602 | 1962 |
| QC Basic Course for Foreman | 6 | 16 | 31882 | 1967 |
| QC Course for Sales Departmant | 13 | 1 | 2059 | 1968 |
| QC Course for Purchasing Department | 16 | 1 | 1726 | 1971 |
| QC Correspondence Course | 6 | 2 | 98749 | 1971 |
| QC Circle Instructor Course | 6 | 19 | 18461 | 1972 |
| QC Circle Top Management Course | 2 | 3 | 2498 | 1973 |
| QC Basic Course for Group Leaders | 4 | 9 | 6972 | 1974 |
| QC Introductory Course for Sales Department | 4 | 3 | 4326 | 1974 |
| TQC Instructor Course | 6 | 4 | 5062 | 1976 |
| QC Course for GMP (Pharmaceutical) | 3 | 1 | 569 | 1977 |
| QC Introductory Course for GMP (Pharmaceutical) | 2 | 1 | 662 | 1977 |
| QC Circle Leader Course | 3 | 38 | 34642 | 1977 |
| QC Circle Course for Clerical Work | 6 | 7 | 6456 | 1979 |
| QC Circle Middle Management Course | 2 | 2 | 737 | 1980 |
| QC Introductory Course for Executives and Managers | 3 | 4 | 5586 | 1981 |
| QC Introductory Course for Purchasing Department | 4 | 1 | 626 | 1983 |
| QC Introductory Course for Sales Department | 4 | 2 | 1974 | 1983 |
| TQC Introductory Course for Service Industry | 6 | 1 | 295 | 1984 |
| Introductory Course for 7 Management Tools for QC | 3 | 14 | 7077 | 1984 |
| Policy Management Seminar for TQC | 3 | 1 | 83 | 1989 |
| Introductory Course on 7 Management Tools for Sales Department | 4 | 1 | 84 | 1989 |
| Introductory Course for Quality Function Deployment | 4 | 1 | 84 | 1989 |
| RELIABILITY | | | | |
| RE Basic Course | 15 | 2 | 4051 | 1960 |
| RE Introductory Course | 4 | 9 | 7215 | 1965 |
| RE Management Course | 4 | 3 | 2933 | 1966 |
| RE Course on FMEA-FTA | 2 | 14 | 9959 | 1976 |
| RE Course on Design Review | 3 | 11 | 6974 | 1977 |
| RE Six-day Course | 6 | 2 | 1517 | 1980 |
| RE Course on Checklists | 3 | 5 | 396 | 1989 |
| RE Course on Test | 3 | 3 | 1393 | 1983 |
| RE Course on Failure Analysis | 3 | 3 | 1048 | 1985 |

**Table 8.1** JUSE education and training courses

- Ishikawa Prize.
- QC Circle Headquarters.
- Quality month.
- Library service.

The Deming Application Prize has a considerable influence on the development of TQC in Japan. It has now been in operation for some 40 years or so (155 awards have been made) and the evidence collected from TQM study missions to Japan (see Chapter 2) indicates that the Deming Prize criteria have produced an almost standard method of approaching TQC. Compared to the West, there is much less company-to-company variation in the level of understanding of TQC and in the degree of attainment. This has helped to promote a deep understanding of quality improvement and TQC amongst all employees. Rather than argue about the merits of a particular approach, system, method or tool, the Japanese tend to discuss how to apply the approach more vigorously through a common core level of understanding. In the West there is some considerable fragmentation of views regarding the best way of approaching TQM, with each expert and their disciples advocating that their approach is the only one which will work over the long-term. By comparison with the Japanese experience, this fragmentation has been detrimental to the speed at which Western companies have adopted TQM.

**The role of academic institutions**

It is noticeable that a number of the Quality Management Centres in UK educational establishments which are involved in the dissemination of quality-related knowledge have been sponsored by the private sector in response to the initiatives of individual academics, and not by government. The same is true for other European countries. At the present time, these are rather like 'islands of information' in a sea of change.

Educational establishments are a key part of a country's quality improvement infrastructure. Several writers such as Morrison[9] and Followell and Oakland[10] have concluded that UK educational and training facilities for quality management are poor. Our own experience in compiling and co-editing the second edition of the *Directory of Quality Training and Education in the UK* bears this out.[11] We discovered that a number of educational establishments appeared to have problems in understanding what quality-related training and education actually entailed.[12] A number of others believed that quality is a basic technical subject unworthy of inclusion in their management courses. There is a tendency amongst some academics to assume that by including quality-related inputs into their engineering and management courses they are catering for the needs of industry. Furthermore, cuts in funding and staffing levels have had an adverse effect on the provision of quality-related education in more than one academic institution. Analysis of the latest edition of the Directory prepared by the Institute of Quality Assurance[13] indicates that the situation has not noticeably improved. We note with interest that many of those educational establishments which provided a negative response in 1985 to our questionnaire survey about their offerings in quality-related training now claim considerable expertise in the subject of TQM.

We agree with Oakland and Grayson[14] that there is a clear need to co-ordinate the content and objectives of quality-related education and training output in the UK to ensure that it does meet the needs of industry, and to fill the gaps in its provision. This is a key role that could be played by a British version of JUSE to ensure that the training and educational resources of academic establishments, professional institutions, etc. make an effective contribution to the technology transfer infrastructure.

**Quality clubs**

The development of Quality Clubs can add impetus at the local 'grass-roots' level. A local Quality Club could be a means of bringing people together from different sections of the local community for the discussion of quality-related topics and for mutual benefit. Such a Club could act as a focal point for the exchange of ideas, education and motivation; perhaps encouraging locally inspired meaningful quality improvement action. Freely available advice on quality-related matters would be an essential element of a Quality Club. All too often small firms do not take advantage of schemes such as the DTI's Enterprise Initiative, because they perceive no need for a formal consultancy service. What such firms do need, which is usually lacking, is a central contact point through which they can tap into expert advice. In short, they often just want to pick someone's brain on an informal basis as and when problems occur. At the same time there is a rich vein of expert knowledge. A Quality Club could bring together people in an informal way to their mutual benefit. Quality Clubs could form a useful adjunct to local enterprise agencies. There are one or two such clubs and forums, e.g., Trafford Park

Performance and Quality Forum and the Quality North Campaign, already in existence.

### Quality technology centres

Major purchasers in the private sector should be encouraged to organise their resources into centres for quality technology. These centres could be government approved (on the same lines as British Calibration Service test houses), offer a comprehensive range of services (e.g., quality and reliability engineering, metrology, quality cost analysis, quality training facilities), and be co-ordinated by a British version of JUSE. These centres would not only complement the consultancy services already available, but could also act as a convenient mechanism with which to develop the local vendor community.

### Q-Share

Typically, small companies have limited working capital, a relatively small 'head-count' of employees and managers, and are forced to be generalists rather than functional specialists. Such companies usually cannot afford to employ an experienced full-time quality manager, and consequently have limited quality management expertise. One remedy is for experienced quality managers to be shared by a group of small companies. Rather than acting merely as consultants, these peripatetic managers can play only a functional role in the companies they serve – preparing quality manuals, conducting supplier audits, assisting with the introduction of quality management tools and techniques, assisting with the process of organisational quality improvement strategies. Such a scheme, called Q-Share (a registered service mark of Polyfield Services Ltd., Hatfield, UK), has been running very successfully in the South-east of England since 1984.[15,16]

Through the Q-Share scheme, small companies receive the services of an experienced quality manager, paying pro-rata for the time he/she spends with them, acquiring the kind of professional services and immediate back-up that would be beyond their normal resources. By necessity, Q-share is a local service; client companies must be within a 40 mile radius of the quality manager's base in order that they may receive the immediate service they need.

## The European dimension

European business organisations tend to import quality management expertise and experience, using consultants and courses from Japan and the US, but the need to look outside Europe for inspiration and help is diminishing. Governments are taking the initiative to develop national quality campaigns; those of Italy, Spain and the Netherlands provide noteworthy examples. Like the UK's DTI, they are recognising the nature of the challenge and their role in meeting it. The Commission of the European Community has publicly endorsed the TQM cause and is backing quality improvement initiatives with cash. EC programmes such as COMETT, SPRINT and BRITE/EURAM are launching a wide range of TQM-related educational, development, research and technology transfer projects involving companies, academic institutions and consultants. Some universities and business schools are becoming increasingly committed to quality management education and training, but they are still in a minority. However, European managers themselves are demanding total quality professionalism, thus creating a rapidly growing market for education and support services in TQM.

To quote K van Ham:

> *Europe is developing its own way, based on its cultural heritage, government support, professional organisations and education. This movement is led by industry.*[17]

Created in 1988, the European Foundation for Quality Management (EFQM) is in the vanguard of this movement.

The inspiration for the EFQM came from the then President of Philips, Cornelius van der Klugt. He foresaw that the post-1992 'open market' in Europe would attract even greater competitive pressure from North America and the Pacific Rim. This he believed had to be countered in a pan-European way, overriding vested national interests and intra-community rivalries. The belief that in spite of years of propaganda, volumes of writings and numerous declarations of faith, European companies still lag far behind US and, especially, Japanese competitors is shared by many top European executives. Their nightmare of a Single European Market dominated by US and Japanese multinationals stimulated action to redress the balance.

In September 1988, van der Klugt and the chief executives of 13 other large European companies signed a letter of intent in the presence of European Commission president, Jacques Delors, to establish the EFQM as a strong driving force to enhance the competitive position of European companies in world markets. These 14 chief executives wished to combine the experience, resources and forces of their, and other, organisations to create conditions for making quality Europe's highest business objective. To achieve this vision, two missions were proposed for EFQM.[18]

1. To support the *management* of European companies in accelerating the process of making quality a decisive influence for achieving global competitive advantage.
2. To stimulate and, where necessary, assist all segments of the European *community* to participate in quality improvement activities and to enhance quality culture.

EFQM's 14 founding members all rank among Europe's most important companies: Philips, Bull, Fiat, KLM, Electrolux, Volkswagen, Nestle, British Telecom, Olivetti, Dassault, Bosch, Ciba-Geigy, Renault and Sulzer. By the end of 1992, membership of the foundation had grown to over 200 organisations. These include European companies of similar size and prestige to the founding members, e.g., ABB, Banco Bilbao Vizcaya, Daf, ICI, Pirelli and Volvo. A number of the European subsidiaries of American companies which have demonstrated a long-term economic commitment to Europe, such as 3M, American Express, Kodak, Exxon, Du Pont and IBM, have been accepted as members. In addition, a growing number of universities and business schools have joined the foundation.

The EFQM is a management organisation. It does not seek to duplicate or usurp the role of existing organisations of quality professionals such as the European Organisation for Quality (EOQ). Indeed, EOQ is an important collaborator in the EFQM's activities – its members' experience is needed by senior managers to support quality improvement strategies and implementation plans. Rather, the EFQM is a joint initiative by top managers to speed up the Total Quality movement in Europe by establishing a process of common learning and sharing of information. This is what makes the EFQM different; the willingness of 'captains of business' from different countries to use the economic muscle of their companies to create a momentum for change on a pan-European scale. This is a critical development in the slow process of raising TQM awareness and changing attitudes and behaviour. Through the EFQM a network of powerful and demanding change agents has started to emerge. In addition, the EFQM is developing as an important technology transfer network in its own right.

To achieve its mission the EFQM is focusing many of its activities on two top priority areas: Recognition, and Education, Training and Research.

## Recognition

Recognition of excellence of performance by people and organisations is regarded as a very important component of EFQM policy, because it is realised that achievement rewards provide exposure and stimulate the exchange of knowledge and further improvement. It has been said that:

> *Measurement gets things done, but recognition gets things done better.*[19]

Already the EFQM operates three European Award Schemes. These are:

- The Research Thesis Award for best doctoral thesis and for best masters or post-graduate thesis related to TQM.
- The Media Award for the best journalistic article or broadcast programme dealing with TQM.
- The Leadership in TQM Award for the best manager demonstrating excellence of TQM leadership. (This has now been subsumed under the European Quality Award for Companies.)

To these has been added the European Quality Award for Companies. To receive the Award, a company must be able to demonstrate that its approach to TQM has contributed significantly to satisfying the expectations of customers, employees and others with an interest in the company.

Applications for the European Quality Award are assessed on the following criteria:

- *Customer satisfaction (20%)* – the perceptions of external customers, direct and indirect, of the company, its products and services (Reasoning: the main stay of good business results).

- *People (18%)* – people's feelings about the company, and the management of the company's people (Reasoning: everything gets done through people).

- *Business results (15%)* – the company's achievement in relation to its planned business performance (Reasoning: the main reason for a company's existence).

- *Processes (14%)* – the management of all the value-adding activities within the company (Reasoning: how people are harnessed to deliver results).

- *Leadership (10%)* – the behaviour of all managers in moving the company towards Total Quality (Reasoning: nothing gets done or improved without effective leadership, policies and strategies).

- *Resources (9%)* – the management, utilisation and preservation of financial resources, information resources, and technological resources (Reasoning: people make things happen; other resources make things possible).

- *Policy and strategy (8%)* – the company's vision, values and direction, and the ways in which it achieves them (Reasoning: see Leadership).

- *Impact on society (6%)* – the perceptions of the community at large, relating to the company. Views on the company's approach to quality of life, to the environment and to the need for preservation of global resources are included (Reasoning: companies as well as people must have respect for and conform to the core values of society at large).

The criteria are split into two groups: 'Enablers' and 'Results'. The European (TQM) model is based on the principle that processes are the means by which the organisation harnesses and releases the talents of its people to produce Results. In other words, the processes and the people are the Enablers which provide the Results. This is illustrated by Figure 8.4.

Essentially, the model shows that Customer Satisfaction, People (employee) Satisfaction and Impact on Society are achieved through:

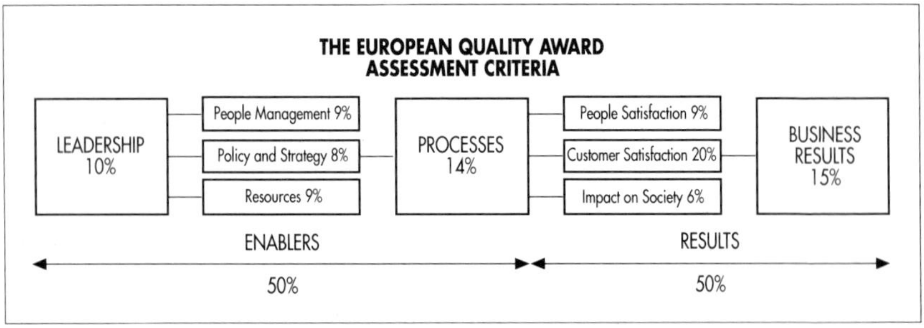


**Figure 8.4** The European Quality Award Criteria Model

- Leadership driving . . .
- Policy and Strategy, People Management, Resources and Processes . . .
- Leading ultimately to excellence in Business results.

Each of the model's nine elements is a criterion which can be used to appraise the organisation's progress toward TQM.

The Results aspects are concerned with *what* the organisation has achieved. The Enablers aspects are concerned with *how* results are being achieved.

The first Award was announced and presented during the Fourth European Quality Management Forum in Madrid in October 1992. The Award for 1992 incorporates:

- *The European Quality Prize* – awarded to a number of companies that demonstrate excellence in the management of quality as their fundamental process for continuous improvement (BOC Limited, Special Gases; Industrias del Ubierna S.A. – UBISA; Milliken European Division; Rank Xerox Limited).

- *The European Quality Award* – awarded to the most successful exponent of TQM in Europe. The trophy will be engraved and held nominally for one year by the recipient (Rank Xerox Limited).

The development of a European TQM model will be the cornerstone of the EFQM's first mission: supporting the management of European companies. The model is intended to help managers of European organisations to better understand best practises and to support them in their leadership role. Already, companies are using it internally to review and guide their TQM initiatives, and are finding it particularly useful in conducting their own self-appraisals.

**Educational, Training and Research**

The Education, Training and Research programmes of the EFQM are intended to help alter the educational infrastructure in Europe, in order to change the attitudes of current and future generations towards recognition of Total Quality as the foremost managerial concept for business. To help realise this ambitious long-term goal, the EFQM is engaged in creating an alliance of educational establishments, companies, national governments and the European Commission. This will be the cornerstone of the EFQM's second mission – assisting all segments of the European Community to make quality Europe's number one value.

The main focal point is a pan-European university-enterprise education, training and research network, established with support from the EC COMETT programme. The network will build on and create partnerships between universities, business schools and companies.

The essential features of the network are:

- Its members will be 'Centres of Competence' in TQM.
- They will be committed to developing TQM, and prepared to allocate resources to this task.
- They are prepared to work in the network in a co-operative and sharing manner.
- They are capable of working at a pan-European level.

Some of the activities undertaken through the network include:

- Collaborative projects between companies and universities to develop knowledge.
- Designing a core curriculum for TQM which would be acceptable and usable anywhere in Europe.
- Joint schemes for the exchange of students, and for the exchange of executives/academics between companies and educational institutes.
- Preparing and distributing new educational materials.
- Guided study visits and technology transfer schemes.

The range of potential opportunities is immense. Whilst the network will cover the whole of Western Europe (including Scandinavia) it will not comprise all educational institutes directly. That would be practically impossible because of the numbers involved. The networks will focus on specific and high priority objectives. It is expected that the members of the network will act as transfer points to national and regional networks.

In the same way, EFQM member companies are beginning to initiate TQM activities in their own regions. The resulting cascade approach is essential for reaching and supporting small and medium-sized enterprises. Member companies, educational establishments, professional organisations, etc. are developing programmes of local EFQM activities. Through networking, it may be possible in time to introduce the Total Quality Culture to all segments of the European infrastructure and to achieve quality as Europe's number one value.

## The American experience

Two of the most famous internationally celebrated gurus, Deming[20] and Juran,[21] were advocating the importance of quality management as a competitive weapon as long ago as the early 1950s. But in what was basically the seller's market of the post-war world their message went largely unheeded in their own country, the USA.

It was probably not until a now famous television programme 'If Japan can why can't we?' featuring Deming was screened in America in 1980 that quality became an issue to be addressed by American industry. Against a background of increasing Japanese automotive imports into the USA in the early 1980s and the impact on employment in the heartland of Middle America, US government and industry leaders became worried about their nation's ability to increase productivity and compete in world markets. The resulting studies of productivity and quality commissioned by the US government and special interest bodies such as the American Society for Quality Control (ASQC), the National Advisory Council for Quality (NACQ) and the American Productivity and Quality Center (APQC) provided the impetus for the creation of a National Quality Campaign with a National Quality Award as its focal point. (For an overview of the relevant events see DeCarlo and Sterett.[22]) The net effect has been to raise the profile of quality in Corporate America. The activities of the American Society for Quality Control received fresh impetus, and a whole new quality management consultancy industry developed (the National Quality Campaign has had a similar impact on consultants in the UK).

Perhaps the most significant act of awareness of the implications of TQM in America has been the creation of the Malcolm Baldrige National Quality Award.

### The Malcolm Baldrige National Quality Award

The Malcolm Baldrige National Quality Improvement Act of 1987, signed by President Reagan on 20th August 1987, established an annual US quality award. The award is named after a former American Secretary of Commerce in the Reagan Administration, Malcolm Baldrige. Its purpose is to promote quality awareness, recognise the quality achievements of American companies, and publicise successful quality management and improvement strategies. The US Department of Commerce and the National Institute of Standards and Technology are responsible for administering the Award scheme.

Up to two awards can be given each year in each of three categories: manufacturing companies or subsidiaries, service companies or subsidiaries, and small business (defined as independently owned, and with not more than five hundred employees). The recipients receive a specially designed gold-plated medal set in crystal. They may publicise and advertise their awards provided they agree to share information about their successful quality management and improvement strategies with other American organisations.

In 1988, 66 companies applied for the Award. Motorola and the Commercial Nuclear Fuel Division of Westinghouse Electric were selected from 45 manufacturing entries, and Globe Metallurgical was the one winner from the 12 small business entries. No winners were selected from the nine service companies that entered.

In 1989, 40 companies applied. Milliken & Company and Xerox Business Products and Systems were selected from 23 manufacturing entries. No winners were selected from the 11 small business entries or from the six service entries.

In 1990, 97 companies submitted application forms and the examiners found that only 12 firms met the awards criteria and from these the judges selected four winners: Federal Express Corporation in the service category (the first time a service company won the award), IBM Rochester and General Motors' Cadillac Motor Car Division in the manufacturing section, and Wallace Company from the small business sector.

In 1991, applications were received from 106 organisations and there were three winners (all electronic firms): Solectron Corporation and Zytec Corporation in the manufacturing section and Marlow Industries in the small business category.

In 1992, there were two winners in the manufacturing category: AT&T Network Systems and Texas Instruments Defense Systems and Electronics Group; two winners in the service category: AT&T Universal Card Services and the Ritz-Carlton Hotel Company; and one small business winner: Granite Rock Company.

Every Baldrige Award application is evaluated in seven major categories with a maximum total score of 1,000. These are: leadership, information and analysis, strategic quality analysis, human resource utilisation, quality assurance of products and services, quality results, and customer satisfaction.[23] The relative scoring weights and relationships of these seven basic categories are shown in Figure 8.5. An outline of the examination categories, sub-categories and scoring weights for the 1992 Award are given in Table 8.2.

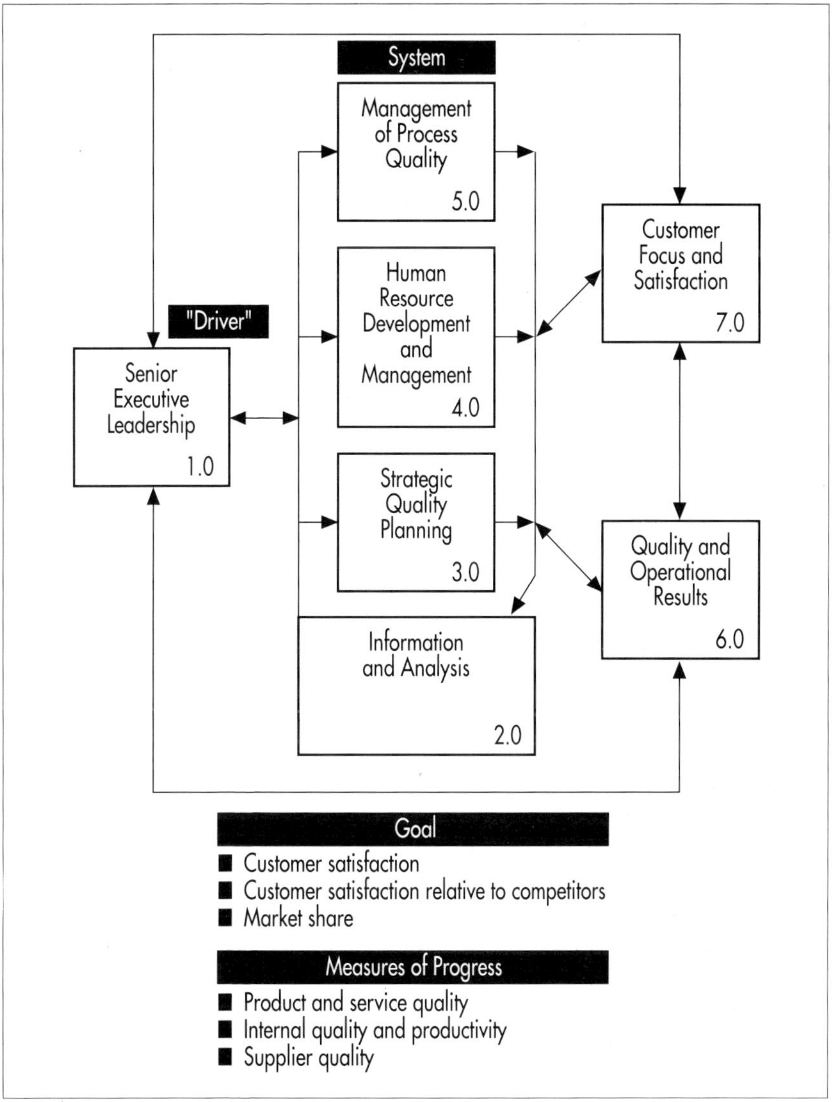


**Figure 8.5** The Malcolm Baldrige National Quality Criteria Model

Entries for the Baldrige Award are scored by more than 175 examiners drawn from industry, government agencies, the military and academia. During the evaluation examiners look for three major indicators of success:

| 1992 Examination Categories/Item | | Point Values |
|---|---|---|
| 1.0 LEADERSHIP | | 90 |
| 1.1 Senior Executive Leadership | 45 | |
| 1.2 Management for Quality | 25 | |
| 1.3 Public Responsibility | 20 | |
| 2.0 INFORMATION AND ANALYSIS | | 80 |
| 2.1 Scope and Management of Quality Data and Performance Data and Information | 15 | |
| 2.2 Completitive Comparisons and Benchmarks | 25 | |
| 2.3 Analysis and Uses of Company-Level Data | 40 | |
| 3.0 STRATEGIC QUALITY PLANNING | | 60 |
| 3.1 Strategic Quality and Company Perforrmance Planning Process | 35 | |
| 3.2 Quality and Performance Plans | 25 | |
| 4.0 HUMAN RESOURCE DEVELOPMENT AND MANAGEMENT | | 150 |
| 4.1 Human Resource Management | 20 | |
| 4.2 Employee Involvement | 40 | |
| 4.3 Employee Education and Training | 40 | |
| 4.4 Employee Performance and Recognition | 25 | |
| 4.5 Employee Well-Being and Morale | 25 | |
| 5.0 MANAGEMENT OF PROCESS QUALITY | | 140 |
| 5.1 Design and Introduction of Quality Products and Services | 40 | |
| 5.2 Process Management – Product and Service Production and Delivery Processes | 35 | |
| 5.3 Process Management – Business Processes and Support Services | 30 | |
| 5.4 Supplier Quality | 20 | |
| 5.5 Quality Assessment | 15 | |
| 6.0 QUALITY AND OPERATIONAL RESULTS | | 180 |
| 6.1 Product and Service Quality Results | 75 | |
| 6.2 Company Operational Results | 45 | |
| 6.3 Business Process and Support Service Results | 25 | |
| 6.4 Supplier Quality Results | 35 | |
| 7.0 CUSTOMER FOCUS AND SATISFACTION | | 300 |
| 7.1 Customer Relationship Management | 65 | |
| 7.2 Commitment to Customers | 15 | |
| 7.3 Customer Satisfaction Determination | 35 | |
| 7.4 Customer Satisfaction Results | 75 | |
| 7.5 Customer Satisfaction Comparison | 75 | |
| 7.6 Future Requirements and Expectations of Customers | 35 | |
| TOTAL POINTS | | 1000 |

Source: Malcolm Baldrige National Quality Award: Gaithersburg, USA

**Table 8.2** Malcolm Baldrige National Quality Award Criteria

*Approach* – What strategy and methodology does the organisation use in attempting to achieve world-class quality?
*Deployment* – What resources are being applied, and how widespread is the quality effort throughout the organisation?
*Results* – Is there convincing evidence of sustained improvement?

Most entrants tend to score within a fairly narrow range on 'approach', and a few fall down on 'deployment', but it is 'results' that separates the real contenders from the rest. High scoring on 'results', which are heavily weighted towards customer satisfaction, requires convincing data that demonstrates steady improvement over time, both internally and externally. Experience has shown that, even with a good internal approach and deployment strategy, it takes time for results to show, so theoretically no company can win simply by submitting a slick entry. Finalists are subjected to 350-400 hours of evaluation by an average 16 expert examiners, including four-day site visits. Quantitative results weigh heavily in the judging process, so applicants must be able to prove that their quality efforts have resulted in sustained improvements. The thoroughness of the judging process means that even applicants not selected as finalists get valuable feedback on their strengths and weaknesses.

US companies often assess their chances of winning before actually entering the competition by conducting an internal audit using the current application guidelines and evaluation criteria. Some large companies rate their chances of success by holding inter-divisional competitions based on the Baldrige evaluation criteria, and nominating the winner as the Baldrige entry.

There is no doubt that the Malcolm Baldrige National Quality Award has helped to raise the profile of Total Quality in the USA. Increasingly, the Baldrige guidelines and criteria are being seen as a TQM role-model and a powerful tool for self-assessment. In 1990, US companies requested over 100,000 copies of that year's guidelines, compared with 65,000 in 1989 and only 12,000 in 1988. Moreover, Baldrige Award examiners, judges, administrators and past winners made more than 3,000 formal presentations on the Award to an estimated 100,000 organisations in business, education, government and health-care.[24] American universities and business schools are using the award as a focus for TQM education programmes and seminars.

In fact, more and more US academics are taking an interest in TQM as a result of the publicity surrounding the award. There is also a growing interest in the award among small and medium sized companies, which recognise that a Baldrige entry would give them a powerful edge in competing for business from companies such as Motorola, Westinghouse and Xerox. Motorola has even gone so far as to notify its suppliers that in order to keep doing business with Motorola they must come up with plans and a target date for entering the award competition. In doing so Motorola claims that it cannot reach its own targets of world-class competitiveness until all its suppliers strive for similar standards.

However, there is increasing disquiet in the US about the nature of the Baldrige Award. A growing body of opinion, including such distinguished names as Crosby,[25] is concerned that the award competition is too bureaucratic and has become a matter of style over substance. In the editorial of one US business journal[26] the Baldrige application process has been likened to the preparation of an income tax return, in that it is cumbersome, time-consuming and expensive. Just as dealing with the Inland Revenue has created a good living for tax consultants, so Baldrige is spawning its own army of consultants ready to help companies play all the angles in the Baldrige competition. Critics feel that the public relations hype surrounding the Baldrige Award is encouraging companies to adopt a 'tick the box' approach to winning the competition rather than a true passion for excellence. They complain that some companies put more energy into creating an award-winning application than they do into actually making the fundamental changes to achieve the results the award signifies.

This is a debate which will probably take several more years to resolve. No doubt many of the negative comments are justified, although two things are certain:

- The Baldrige Award has placed quality firmly on the agenda in American boardrooms.
- It goes a long way towards giving business, academia, consultants and government a tangible model of Total Quality in action.

## Summary

Based on the experiences of the UK National Quality Campaign, government can most certainly help to increase industry's awareness of the need to improve product and service quality. Such a campaign is likely to have the greatest impact on the small and medium sized company sector. Governments, of course, cannot empower organisations to adopt quality improvement as a key business strategy.

The strategy of the DTI has been consistent throughout the campaign and the current Enterprise Initiative – awareness through brochures, videos, management briefings, workshops and the like and assistance for small companies to develop their quality-related system. The awareness initiatives taken of late in terms of management briefings and videos are somewhat repetitive of early material. It is our view that newer and bolder initiatives should have been taken in the area of technology transfer of quality improvement knowledge and practise which have been outlined in this chapter.

Facilitating technology transfer of quality-related information is not easy, and each person, body, organisation, and institution with an interest in the topic tends to think their methods are best and do their own thing. However, establishing Centres of Excellence and setting up networks and clubs is clearly something which the government can influence and promote.

A challenge facing other governments considering promoting their industries awareness of quality is to provide a consistent message and yet at the same time introduce new initiatives and facilitate technology transfer.

The criteria and factors of Quality Award Schemes which recognise the quality achievements of organisations are invaluable in helping companies to develop the key aspects of their quality improvement process. These awards also help to pinpoint the weaknesses in the current approach and guide the organisation to superior performance. The Deming Application Prize has had a major influence in Japan and the early indications are that the Malcolm Baldrige Quality Award will do the same for American organisations. It is hoped that a European award will do the same thing for the whole of Europe.

## References

1. Lascelles, D. M., and Dale, B. G., 1989, "The UK Department of Trade and Industry National Quality Campaign 1983-89," *International Journal of Operations and Production Management*, (Monograph), **9** (6).

2. Department of Trade and Industry, 1985, *Getting to Grips with Quality*, Quality, Design and Education Division, Department of Trade and Industry, London.

3. BS 5750/ISO 9000, 1987, *Quality Systems*, British Standards Institution, London.

4. Lascelles, D. M., and Dale, B. G., 1989, "The National Quality Campaign: a Study of its Impact on Industry," *Proceedings of the Institution of Mechanical Engineers*, **203**, 201-209.

5. Department of Trade and Industry, 1986, *Quality Counts*, Quality, Design and Education Division, Department of Trade and Industry, London.

6. Department of Trade and Industry, 1992, *Register of Quality Assessed UK Companies*, Department of Trade and Industry, London.
7. Department of Trade and Industry, 1985, *Caring Enterprise and Winning World Market Share*, Quality, Design and Education Division, Department of Trade and Industry, London.
8. Union of Japanese Scientists and Engineers, 1990, *Organisation and Activities*, JUSE, Tokyo.
9. Morrison, S. J., 1981, *International Comparative Quality Assurance: A Pilot Study*, University of Hull.
10. Followell, R. H., and Oakland, J. S., 1985, "Research into Method of Implementing Statistical Quality Control," *Quality Assurance*, **11** (2), 27-31.
11. Lascelles, D. M., and Dale, B. G., 1985, *Directory of Quality Training and Education in the UK*, Second Edition, Quality, Design and Education Division, Department of Trade and Industry, London.
12. Lascelles, D. M., and Dale, B. G., 1985, "Quality Training and Education: A State-of-the-art Survey," *Quality Assurance*, **11** (2), 39-42.
13. Institute of Quality Assurance, 1990, *Directory of Quality Training and Education in the UK*, Institute of Quality Assurance, London.
14. Oakland, J. S., and Grayson, R., 1987, "Quality Assurance Education and Training in the UK," *Quality and Reliability Engineering International*, **3** (3), 169-175.
15. Deshpande, A. B., Dusting, F. P. J., and Younger, A., 1986, "A Co-operative Quality Management System for Small Companies," *International Journal of Quality & Reliability Management*, **3** (1), 38-47.
16. Younger, A., 1990, "Q-Share of Total Quality Management," *Proceedings of the 1st European Conference on Education, Training and Research in Total Quality Management*, IFS Ltd., Bedford, 69-75.
17. Van Ham, K. J., 1989, "Quality Management in Perspective," Management Brochure 1, European Foundation for Quality Management, Eindhoven.
18. European Foundation for Quality Management, 1988, *European Foundation for Quality Management*, European Foundation For Quality Management, Eindhoven, The Netherlands.
19. European Foundation for Quality Management, Peacock.
20. Deming, W. E., 1945, *Statistical Method from the Viewpoint of Quality Control*, Lancaster Press, New York.
21. Juran, J. M., 1945, *Management of Inspection and Quality Control*, Harper and Row, New York.
22. DeCarlo, N. J., and Sterett, W. K., March 1990, "History of the Malcolm Baldrige National Quality Award," *Quality Progress*, 21-27.
23. United States Department of Commerce, 1990, *Application Guidelines, Malcolm Baldrige National Quality Award*, National Institute of Standards and Technology, Gaithersburg, MD, USA.
24. Reimann, C. W., 15th October 1990, "America Unites Behind the Baldrige Quality Crusade," *Electronic Business*, 63.
25. Crosby, P. B., and Reimann, C. W., May 1991, "Criticism and Support for the Baldrige Award," *Quality Progress*, 41-44.
26. McLeod, J., May 1991, "Style, Substance, Baldrige," *Electronics*, 4.

# Chapter 9
# Creating a Total Quality Improvement Strategy

## Introduction

Total Quality initiatives often start and end with a vision and mission statement, and a burst of public relations activity. Unquestionably, the least effective way of starting TQI is for the chief executive suddenly to declare a new found zeal for the philosophy of TQM, declare a corporate vision, initiate a PR-style motivation programme, send people to TQM awareness seminars, form quality improvement teams, and then wait for something to happen. If the chief executive officer is relying on this action alone to stimulate action, he/she will be disappointed. He/she will also be negligent in his/her duty to provide the company with business direction. This kind of hand-on-heart, cliché-laden, quick-fix approach may conceivably result in improvements, but these will be short-lived, leading to the cycle of stagnation, attempted programme renewal and ultimate decline as illustrated in Figure 1.10. Worse still, the concept of Total Quality will be devalued and discredited in the eyes of most of the organisation's members, making any subsequent attempt to introduce Total Quality more difficult, as resistance to change will be more deeply entrenched.

Like any other business strategy, TQI necessitates planning and careful direction by the chief executive and his/her senior management team. We believe that the creation of a strategic framework for the implementation of TQI should take into account the following eight steps:

Step 1 – Condition setting.
Step 2 – Policy formulation.
Step 3 – Policy deployment.
Step 4 – Improvement projects.
Step 5 – Quality system.
Step 6 – Process improvement.

Step 7 – Process stream improvement.
Step 8 – Strategic benchmarking.

Each of these eight steps is now described in more detail.

## Condition setting

By 'condition setting' we mean setting the scene for creating a TQI culture – a culture for organisational change. This is an essential prerequisite to the formulation and execution of a TQI policy and implementation strategy. It has been said that leadership originates from the creation of a vision. However, too many policies and strategies for change are envisioned and articulated by top mangers without serious consideration of the environmental conditions and their consequences for the organisation. The senior management team must focus their attention both internally, to determine the organisation's strategic needs, and externally, to develop their own TQM knowledge base, in order to establish a common sense of mission and direction.

The chief executive and senior management team must enhance their own understanding of TQM by finding out as much as possible from external sources about the philosophy and its underlying principles. This involves reading around the subject, attendance at conferences and seminars, joining 'quality clubs', 'forums' and networks, forming links with academic institutions with TQM education and research track-records, and visiting other companies with reputations for being 'centres of excellence' to learn at first hand the lessons of TQI planning and implementation.

Competitive or comparative benchmarking of excellent systems and product/service offerings generates a stream of useful data which will help facilitate TQI.

Before a coherent TQI strategy can be formulated, it is first necessary to determine the organisation's current status both operationally and culturally. This means assessing the effectiveness of processes and operational systems on the one hand, and people's attitudes on the other, viz., diagnostic surveys of company strengths and weaknesses, employee attitudes, customer satisfaction and cost of quality.

### Strengths and weaknesses

We suggest that an organisational strengths and weaknesses diagnoses should cover the following areas:

but if the costs later appear difficult to influence through direct action, credibility is lost. Definitions are a key feature of quality costing.[2]

When preparing a survey of quality costs it is recommended that all the interested managers get together to agree the 'ground rules', and to reach a common definition of quality costs. This is important if quality costs are to have any credibility as a company-wide tool for measuring and tracking improvement. The outputs of such a workshop should include:

- An understanding of the definition of quality costs, the six category elements, i.e., prevention, appraisal, internal failure, external failure, redundant activity, lost opportunity, and basic work (the essential activities entailed in supplying a company's products and/or services).
- An agreement on a common system for allocating costs to the various elements of quality costs.
- An agreement on the methodology and approach, allocation of resources and timetable for completion of the survey.

The desired outcomes of the quality cost survey include:

- An overall cost of quality figure to be used as a management stimulus as required.
- A management index of quality performance.
- A ranked list, by cost, of potential improvement projects for the main business areas.
- Some guidelines for designing an appropriate system for updating and monitoring quality costs in the future.
- If the quality costing approach has been based on an analysis of activities it lays the foundation not only for departmental purpose analysis and the internal customer/supplier concept, but also for activity-based costing.

**Top-team workshop**

In order to formulate an effective TQI policy and implementation strategy, it is vital for the senior management team to ensure that they themselves have a common understanding of and orientation to the principles and values of Total Quality. In our experience this is best achieved through a short, residential, off-site workshop for the senior management team, perhaps with an external consultant or advisor acting as a facilitator. The purpose of such a top-team workshop is to review and assess the company's current position with regard to its business, customer needs and its management structure. Its outputs should include:

- A common understanding of TQM principles and values.
- A team view of the *status quo* – 'where we are now'.
- A team vision for the next three-to-five and five-to-ten years – 'where we want to be'.
- A means of integrating quality improvement objectives into the organisation's business objectives.
- Identification of possible barriers to the achievement of the vision that will require action by the team.
- Some understanding of the scope and depth of the organisational changes which are necessary.
- Creation of personal action plans.

Naturally, these outputs will need further development and refinement after the workshop, but they form the basis for policy formulation.

**Quality Council**

The next step is to form a steering group or quality council. The overall role of the quality council is to co-ordinate and oversee the TQI strategy. Its key task is to ensure that all reporting managers understand the principles of the organisation's TQI strategy, that they have clear and appropriate quality improvement plans in place, and that they understand how to go about implementing them. It is the quality council's responsibility to identify barriers to improvement which are outside the responsibility of reporting managers, and to assign ownership and responsibility for their removal. Depending on the size of the organisation, there might be just one company-wide quality council or a top-team quality council comprising the most senior managers and a series of satellite councils representing and overseeing the TQI activities of individual business units. The top-team council's role is to co-ordinate the activities of satellite councils ensuring that they correspond to the overall TQI policy, and to provide a structured feedback of the lessons learned from the management of improvement projects so that all parts of the business can share and benefit from the accumulated experience.

The quality council is primarily an engine for organisational change concerned with the shaping of new cultural values, therefore it must be chaired by the chief executive. Likewise, business unit quality councils must be chaired by the unit general manager. By all means have a quality director reporting at Board level as this will help ensure that quality is kept on the Board's agenda. However, the quality director's task is not the same as that of the quality council's chairman. The quality director is responsible for the integrity of the company's quality system and acts as chief quality improvement facilitator or in-house consultant, perhaps as head of the company's quality improvement technical support unit. The quality council chairman's task is to lead the formulation, co-ordination and implementation of a cross-functional business improvement strategy. As we pointed out in Chapter 5, only the chief executive has the necessary authority to do this.

## Policy formulation

### Vision, mission and values

Policy formulation is about setting a direction: developing a vision of the future plus the strategy needed to achieve that vision, describing a business/organisational culture in terms of what it should become in the long-term and articulating a feasible way of getting there. A policy is a basic framework of shared business principles and values that helps to guide the organisation's members in their day-to-day activities towards the vision. So care must be taken that it encompasses a set of realistic business goals on which the organisation's members can focus, and that it is not merely a philosophical statement.

An effective policy addresses four key elements:

*The Vision* – the desired state the organisation wishes to attain; what it wants to be.
*The Mission* – how the organisation is going to get there.
*Goals* – the key improvements needed to achieve the mission.
*Values* – the shared beliefs which guide the organisation's actions in reaching towards the vision.

So a company might, for example, express its vision as: "To become the leading low-cost producer in (our market sector) consistently meeting and exceeding the expectations of our customers."

The mission could be: "To become a totally well-managed company by improving the effectiveness of all our processes company-wide to the extent where we might compete for the Baldrige Award/European Quality Company Award/Deming Application Prize within five years time."

Sir Colin Marshall, chief executive of British Airways, has defined the importance of a corporate mission to the creation of an improvement strategy: "A corporate mission is much more than good intentions and fine ideas. It represents the framework for the entire business, the values which drive the company and the belief that the company has in itself and what it can achieve."[3]

Philips Car Stereo has clearly stated corporate values and beliefs which can serve as a framework for any organisation:

**Philips Car Stereo: Values and Beliefs**

1. *Quality*
   The ultimate measure of quality is the customer's perception of our products and our performance.

2. *Leadership*
   We are committed to leadership in technology and innovation, directed towards extra added value.

3. *Delighting the customer*
   We aim to delight all our customers by constantly delivering performance that surpasses their expectations.

4. *We keep our promises*
   We are a business partner our customers can count on. That means when we make a commitment, we keep it.

5. *We don't hide failures, and we correct our mistakes*
   Perfection is an unachievable goal. But we never stop trying. We work together on solving problems, and we do everything to make sure mistakes don't happen again.

6. *Our efficiency can always be improved*
   However much we achieve, we always try to do better. Because that's the only way to stay ahead of the competition.

7. *If a customer has a problem it's our problem*
   No trouble is too great to solve a customer's problem. Responsive service demands personal initiative and effort.

8. *We aim to create a positive working environment*
   We recognise that our people and their skills are our most important resources. We therefore aim to create a stimulating and rewarding working environment for everyone at Philips Car Stereo.

*If we can achieve these aims, we will have gone a long way towards becoming a Total Quality company. We have made a lot of progress in the last ten years, but we still have a lot to achieve.*[4]

A word of warning: The vision and mission must be seen to be that of the organisation and its people, and not just that of the chief executive officer and senior management team. Also, the progress being made by the organisation to achieve the vision and mission must be visible and measurable. In some organisations, e.g., Grace Dearborn, Widnes, this is measured by means of an employee attitude survey which addresses issues such as: "Are we closer to achieving the vision and mission than we were two years ago?"

Policy formulation will already have begun with the top-team workshop, from which a fundamental Total Quality vision and mission will have emerged. Effective TQI depends on the top management team's ability, either as a board of directors or as the nucleus of a quality council, to integrate this vision and mission with the overall business direction of the company. In formulating a policy, it is useful to develop a TQI model on which everyone in the organisation can focus their attention.

Managers, facilitators and management consultants often operate with only a partial model in mind about why they are doing whatever it is they are doing, and what principles lie behind it. This reminds us of the old story about one Irish prime minister who after a particularly long cabinet debate declared: "I understand how it works in practice. But how does it work in theory?" In formulating an improvement strategy, the underlying theory can be almost as important as the practical application.

Figure 9.1 shows a theoretical model of the basic TQI mission which combines the vertical deployment of policy, to ensure that everyone in the organisation pulls in the same direction towards specific goals, with the horizontal deployment of processes, to align all process interfaces and customer/supplier relationships to increase customer delight. An effective TQI strategy is concerned with linking policy deployment and process deployment to add value and achieve competitive advantage.

This type of model may be used with some adaptation to suit the needs of the organisation's own culture and business environment.

**Goals and objectives**

Policy formulation will involve setting long-term, company-wide goals and converting them into top-level quantifiable objectives that are

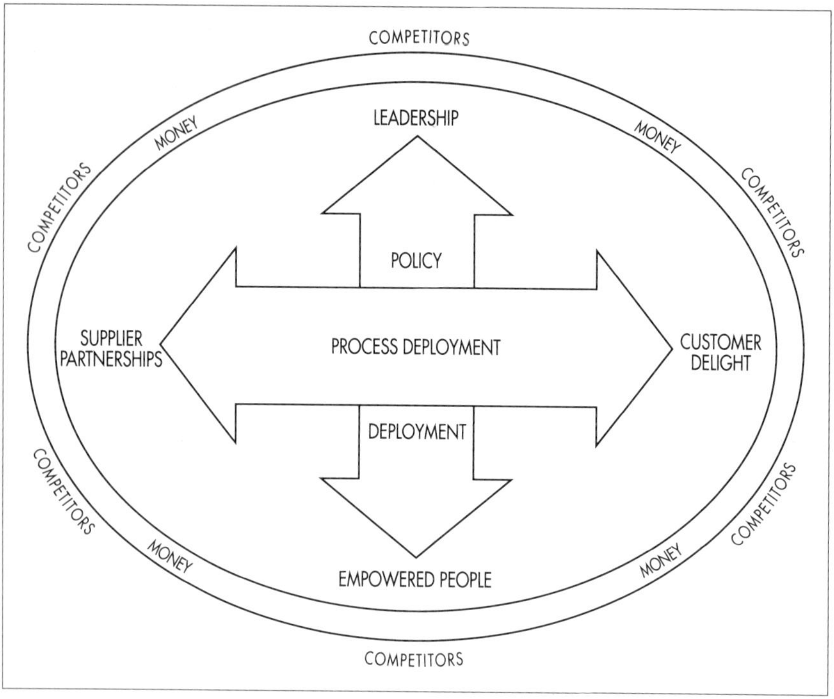


**Figure 9.1** TQI policy model

specific and measurable. Objectives must have an external as well as an internal orientation, otherwise a TQI strategy may achieve its objectives and still not result in improved customer satisfaction. The internal benefits of TQI are largely irrelevant to the customer who is concerned only with how well a product or service satisfies his/her needs and its value for money.

Improvement objectives must also be capable of being used as a tool for management appraisal and motivation. Otherwise, despite good intentions, managers will tend to side-step improvement activities in response to changing priorities. When improvement performance is formally appraised against publicly declared objectives, managers take a positive attitude to improvement activities and prioritise them accordingly. To repeat the oft-quoted phrase: what gets measured gets done.

These objectives, which should be derived from a careful analysis of the business environment carried out as part of the condition setting phase, will form the cornerstone of the company's quality improvement plan. This plan will form the basis for quality improvement plans at business unit/department/section level as it is cascaded downwards through the organisation. Included in this, and the plans generated at each successive cascade level, will be plans for quality improvement projects, communication, education and training, and the application of quality management tools and techniques, plus the assignment of responsibilities for carrying out the various components of the plan(s) and methods of assessing the progress made.

Policy formulation will also involve clarifying short-term priorities, both in terms of the amount and type of resources, e.g., people, financial, information, that can be deployed, and identifying key improvement opportunities. The top management team will have important decisions to make regarding the allocation of what are usually scarce resources and support facilities to areas which will make a significant impact on the overall business plan. A related task will be the identification of core competencies within the organisation (i.e., potential facilitators, team leaders and support personnel) as well as external sources of assistance. They will also have to identify and fix the most appropriate competitive benchmarks, not as performance measures in themselves but as reference points against which to measure progress. Finally, they will have to establish relevant and meaningful performance measures, and a mechanism for auditing achievement against planned objectives.

**Performance measures**

Establishing effective performance measures requires careful consideration of the implications of the policy objectives and their deployment. A genuine understanding of the operational processes and activities affected is an essential prerequisite to the setting of performance measures, therefore this task must be left to the managers who are formulating action plans in response to objectives cascaded downwards. The top management team will of course have to set its own performance measures, which will be tied to top-level action plans and objectives. Whilst engaged in this kind of exercise, more than one top management team has found that its grasp of strategic principles was not as sound as it had initially assumed.

Payne[5] lists the essential elements of good performance measures:

- They must reflect the critical aspects of functional performance – meaningful performance measures depend on a clear understanding of functional purpose.

- They must reflect the critical aspects of performance as seen by the customer – it is fundamental to establish and define the needs of both external and internal customers.
- They must be understood by all those who need to understand – if the improvement process is to be measured and the appropriate people involved, in order to motivate them and make them accountable, then they must understand the measures being used to assess their performance.
- They must be highly visible – since the prime purpose of performance measures is to inform people how well they are doing against set targets, the style of presentation is important.
- They must relate to the requirements of the total business – the local performance measures should have been tested against the objectives of the overall business improvement strategy.
- They must be subject to a reviewing mechanism – in the dynamic environment of modern business change is inevitable and should be monitored to ensure the continuing validity of performance measures.
- They must be used in conjunction with appropriate target-setting – targets, i.e., objectives, define the required rate of improvement, and therefore must be both realistic and challenging.

## Policy deployment

If policy formulation is about setting a direction, then policy deployment is concerned with aligning people to that direction and keeping them moving in the right direction. Policy deployment embodies a wider set of activities than just cascading objectives and priorities down through the organisation. It is, in effect, the process of internalising continuous improvement and the values of TQM throughout the organisation. Furthermore, it is not a one-off activity carried out at prescribed periodic intervals, but a continuing cycle of stimulating action, review, analysis of events and renewal.

### Aligning people

Aligning people involves the chief executive and the top management team communicating the new direction by cascading the vision, goals and priorities down through the various levels of the organisation, thereby helping to create a collective sense of purpose. This involves more than just sharing the vision with all the organisation's members. It necessitates establishing a structured specific and action-oriented objectives as they are cascaded downwards. As the vision and the top-level goals will be expressed as a general statement of direction for change, they will need to be defined as quantified objectives and prioritised as an interrelated set of action plans and performance measures that can be deployed at the 'coal-face' of the business (Figure 9.2).

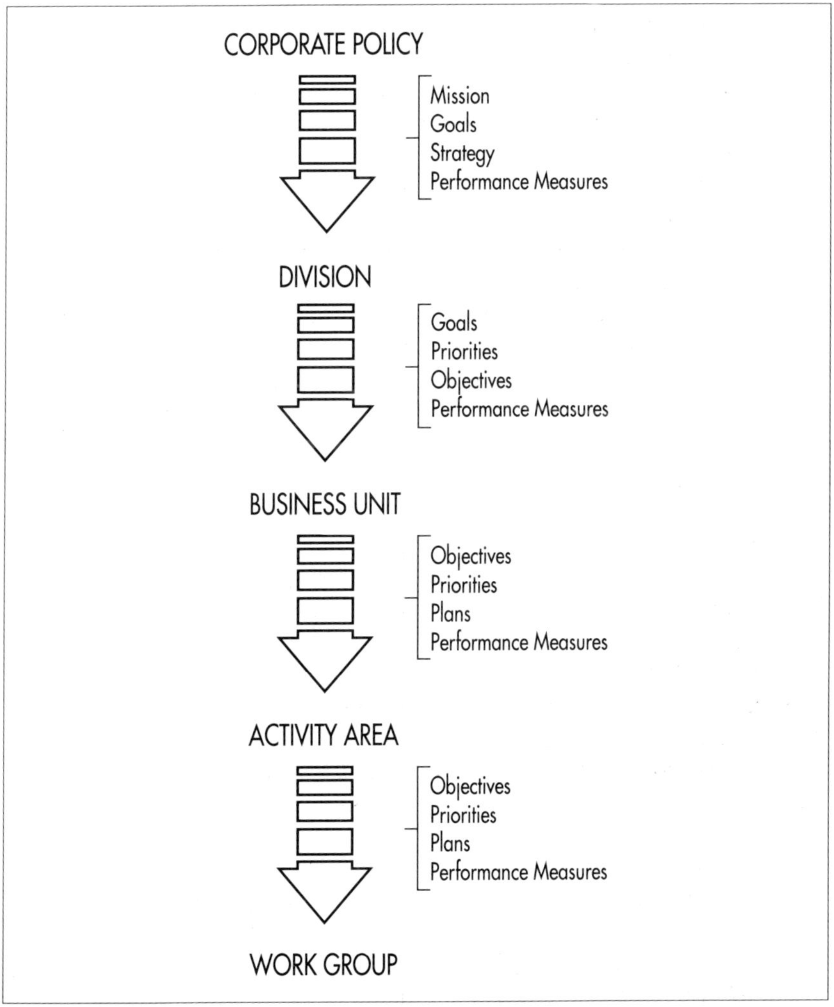


**Figure 9.2** Policy deployment cascade

At each level of deployment it is essential not only that there is a clear understanding of the role of each manager and the business unit/department/section for which he/she is responsible, but also that each manager participates fully in the deployment exercise and is able to accept ownership of the objectives and performance measures set. This is crucial, because whilst the direction of the deployment cascade is top-down, all levels, through their commitment and their actions, will contribute to the effectiveness of the continuous improvement process and, with time, influence the overall strategy. The policy deployment structure must have an in-built closed-loop feedback mechanism through all managerial levels so that the chief executive and the top management team can hear the 'voice of the organisation' to enable them to monitor the effectiveness of the TQI strategy properly, and make changes where necessary (Figure 9.3).

### Education and training

Having aligned people in the direction for change, the next task is to empower them by providing the opportunities, tools and skills to enable them to contribute to the realisation of the goals set.

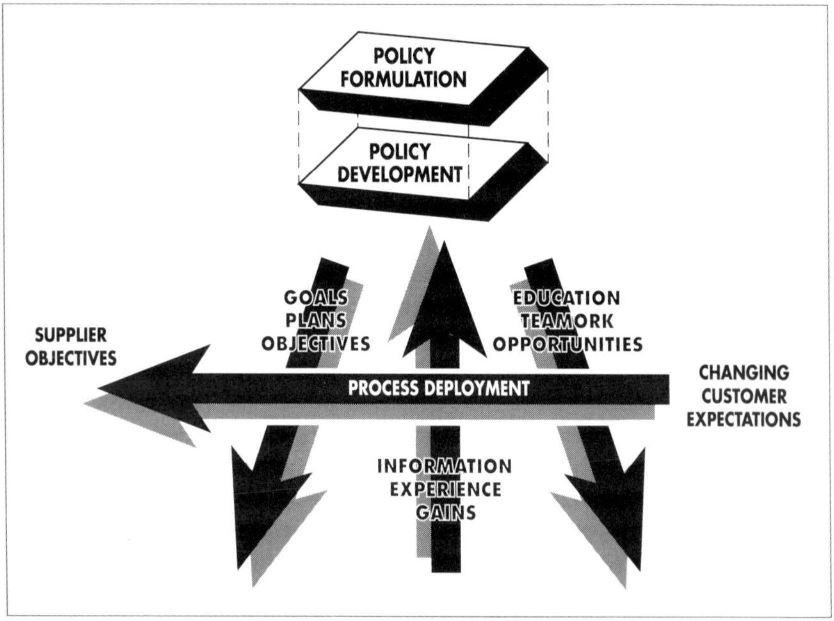


**Figure 9.3** Policy cascade and alignment

Education and training is quite simply a generator of cultural change. It must start at the top with the chief executive and the senior management team. We have discussed some of the barriers to cultural change, e.g., inadequate leadership, fear, inadequate skills, poor information, earlier in the book. These barriers have to be encountered and removed if managers throughout the organisation are to change their attitude and operational style. These lessons are being slowly absorbed by those companies already well down the TQI road. Many such companies have developed in-company training schemes covering, for example, communication skills and techniques, teamwork, problem-solving methodology, recognition and motivation skills, etc.

There is a growing range of approaches to training in Total Quality. These include:

- Awareness sessions.
- Personal focus and commitment workshops.
- Training for trainers, facilitators and improvement team leaders.
- Training for quality support managers and personnel.
- Training for all employees in improvement tools and techniques.

There is now a wide range of training materials, such as textbooks, tutor packs, computer-aided learning packages, distance learning packages, etc.

An increasing number of organisations now view training as an essential strategy to develop an organisational culture committed to delighting the customer, and in which everyone knows his/her job and what he/she has to do to succeed in it. These companies do not see training as just another add-on overhead, to be fitted in when other operational pressures permit the release of employees, or as a candidate for trimming at the earliest opportunity.

Team-based, manager-led workshops and training sessions are the most effective mechanism for deploying education and training. By addressing real problems and needs, training is made directly relevant to the work situation. By involving managers to lead the educational process at all levels, training is still aligned to overall corporate goals (Figure 9.4).

- At each level managers learn from their bosses.
- With an enhanced personal understanding of concepts, apply the knowledge in their own operating environment to help develop objectives, performance measures and deployment plans.

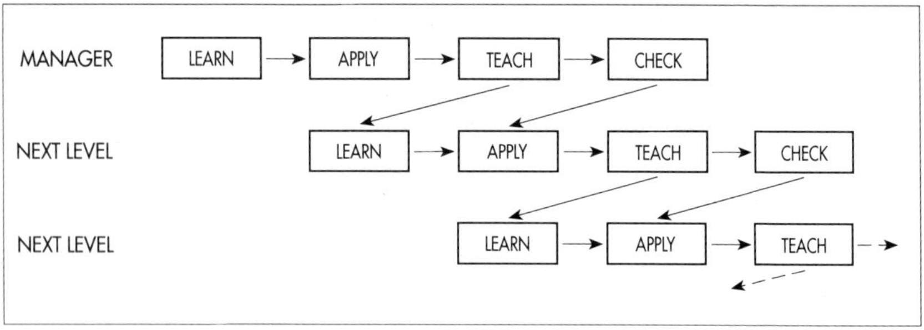


**Figure 9.4** Deployment of education and training

- Pass on the knowledge and skills necessary for teaching their own subordinates.
- Check the effectiveness of the dissemination process by monitoring progress against agreed performance measures.

Education and training should never be carried out just because it is a worthy exercise or to simply improve employee morale; training programmes must always be focused on a specific business-related objective. In our experience, companies rarely set improvement objectives for the training undertaken. The result is wasteful utilisation of resources and the feeling amongst managers that training is a necessary evil – or to paraphrase the first Lord Leverhulme: "Half of my training budget is wasted, but I don't know which half." Training audits to identify how effectively trainees are using their newly taught skills are an essential part of any education and training strategy.

## Motivating people

Enlightened management teams are beginning to realise that one cannot keep people moving in the right direction by decree – fear and effective leadership do not mix. People can only be encouraged to follow a path by motivation and inspiration, by appealing to basic but often untapped human needs, values and emotions. For the chief executive and top-level managers this means a willingness to promote all aspects of Total Quality continuously, both internally and externally, by demonstrating personal commitment and involvement, and motivating others by recognising achievements. This also means encouraging all managers to do likewise, to become leaders of change (or 'change agents').

Change agents, from the chief executive down, must demonstrate their commitment and involvement by their attitude, behaviour and decisions on an on-going basis. This is easier said than done and requires great self-discipline before it becomes habit-forming. Moreover, a change agent must be accessible to fellow-members of the organisation (especially his/her subordinates), customers and suppliers. This means:

- Encouraging questions and discussion.
- Listening to new ideas.
- Giving recognition where it is due.
- Finding time to spend in walking around operating and office areas.

All forms of ad hoc communication which are as much a part of the continuing cycle of policy deployment as the formal cascade and feedback mechanisms.

Personal commitment is also demonstrated and reinforced by participation in Total Quality activities, such as:

- Identifying non-added value activities.
- Taking part in improvement projects.
- Visiting customers and suppliers.
- Dealing with customer complaints.
- Accepting ownership of key improvement processes.
- Leading education and training activities.

Recognition is a primary motivation force. As we stated in Chapter 5, good leaders will help people to grow and find ways to enhance the self-esteem of their colleagues. Recognising and rewarding effort and sharing successes will give people a sense of achievement and reinforce their sense of belonging. This means not only showing appreciation internally to groups and individuals, but also acknowledging the performance of external suppliers. This may also mean changing management and staff appraisal schemes to take account of their performance and achievements in the light of Total Quality principles and values. To quote Anne Van't Haaff (former corporate quality manager, KLM): "What gets rewarded gets repeated."[6]

Effective policy deployment is, in our view, a function of leadership, personal commitment and communication used in combination to create and reinforce a collective sense of purpose. It is no less than the deployment of an organisational truth.

## Improvement projects

Initiating the actual improvement process necessitates the creation of an improvement infrastructure in which the organisation's members can gain experience in problem-solving and participate in change activities. The first quality improvement teams should be formed at the most senior levels of the organisation, and then at progressively lower levels until everyone is involved, or at least has been given the opportunity to get involved. Because most problems will be inter-functional, the team approach is essential, and this helps all participants to gain teamwork experience which will prove invaluable when the focus of the TQI implementation strategy moves to cross-functional process streams.

The prime reason for this exercise is to improve both management skills and experience, particularly quality management skills, by developing a reservoir of problem-solving and project management expertise. Quality improvement projects can help to develop the kind of leadership culture we envisioned in Chapter 5 by providing opportunities for leading change and developing teams. However, improvement project activities are not merely a learning exercise, but should form part of the annual improvement plan and be co-ordinated with the development of a quality system which, in turn, will provide the relevant framework and basis for standardisation.

An extensive training programme will be required to provide people at all levels of the organisation with the problem-solving and teamwork skills necessary to enable them to make a real contribution to the company-wide improvement effort. Everyone should receive training in the use of the 'Seven QC Tools' for analytical problem-solving, and in data presentation skills.

### Visibility and teamwork

By formalising a problem into a project it is given a tangibility through which it can claim the attention, resources and facilities necessary for its solution. A quality improvement project has certain generally recognisable characteristics:

- A specific, clearly defined problem.
- Clear, measurable objectives.
- Better managed by a team than an individual.
- Clear starting and finishing points.

- Capable of being solved through the application of a structured project management approach.
- Potential to contribute to bottom-line profits.

Visibility and teamwork are two core factors which shape the nature of a quality improvement project. A project should focus on a problem visible to the people involved, which is relatively easy to solve, and from which the results of improvement should show in a time frame of 6-12 months. The project must have clear objectives that are measurable and quantifiable, so that the results are visible. It should also be something which can be tackled by a team that will permanently resolve a problem within their own operational environment.

Every project must have a project owner, usually a manager willing to sponsor (to 'champion') one or more quality improvement teams in his or her own operational area. Project ownership is a real leadership activity concerned with organisational change, and carries several crucial responsibilities which will affect the success of the project. These responsibilities include:

- Selecting appropriate projects and measurement criteria.
- Selecting the quality improvement team and its leader.
- Ensuring that the quality improvement team receives the appropriate education and training.
- Reviewing the progress of the project.
- Removing any barriers which inhibit progress, getting support from more senior managers if necessary.
- Communicating and rewarding success, including the recognition of individual contributions to the overall team effort.

**Improvement facilitators**

Project champions and team leaders will need support from improvement facilitators. Facilitators are a key element of the improvement infrastructure, acting as a source of expertise to be tapped by everybody in the organisation from top to bottom. Reporting to the quality council, the facilitator is responsible for facilitation of the TQI process (literally, making it easier for it to happen). He/she is not solely accountable for its success; ownership and accountability rests with the project champions and team leaders.

The facilitator has several roles:

- Giving advice on the *planning* of the process. This may include helping to develop the initial deployment plan and modifying the plan as experience dictates.
- *Identification* of problems, roadblocks, lack of expertise or any other potential cause likely to hinder the smooth implementation of the quality improvement process.
- Acting as the initial *independent interrogator* of internal customer perception and using this and the other inputs (business objectives, quality costs) to help line management to develop and prioritise improvement projects which will form the basis of their improvement plan.
- Acting as an *internal consultant.*
- *Training.* This includes the ability to identify specific training needs and where appropriate deliver such training as team building, project management skills, the seven tools, and so on.
- *Monitoring* the progress and supporting the improvement projects – this is the ideal way for the facilitator to recognise training needs and to ensure that the training is both economical and effective.
- Developing a *facilitator network.*

Ideally drawn from middle management, he/she should:

- Be recognised both by upper management and his/her peers to be highly competent and credible in his/her present occupation.
- Have good inter-personal skills.
- Be a good communicator.
- Be reasonably numerate.
- Be a team player.
- Have good knowledge of the organisation's process network.
- Believe in the participative ethic.
- Be self secure.

He/she should *not*:

- Happen to be 'spare', awaiting redundancy, a casualty of reorganisation or a low performer who is difficult to place.

The number of facilitators required will depend on the size of the organisation.

**Identifying projects**

The quality council has a key role in the project review process. This includes:

- Agreeing and supporting project objectives and implementation plans.
- Helping to run a project by smoothing the quality improvement team's path, eliminating obstacles to progress which lie outside the team's control.
- Monitoring project costs and savings.
- Facilitating the spread of the successes and lessons derived from the project throughout the organisation to maximise the benefits to the business overall.

Juran's[7] differentiation of the most serious and important problems that will provide significant breakthroughs to new and higher levels of performance (what he calls the 'vital few') from the less significant multitude of everyday improvement opportunities (the 'useful many') can be used as a guide to help the quality council and improvement project champions to develop priorities for action. Every business unit within the company should identify its own 'vital few' problems and schedule them for special attention and early solution.

'Vital few' projects can be identified on the basis of one or more such factors as:

- The value of their failure costs (the amount of non-value added or waste).
- Impact on customer satisfaction (the results of a customer satisfaction survey will yield important clues).
- Ability to inhibit other significant improvement plans or projects.
- Ability to contribute in some important way to achieving the corporate vision.

They are likely to be cross-functional, will need high-level ownership (at board level) and assistance in removing obstacles, and may well be long-established problems which, in the past, have resisted solution. They are leadership opportunities requiring the involvement of a senior champion. By their nature they are longer-term, senior management driven, top-down projects which have important resource and priority implications, not least of which will be to ensure that day-to-day operational issues do not impede progress.

The 'useful many' projects, on the other hand, will bring about significant overall improvements because of their sheer number. These will primarily be:

- Intra-department project opportunities within the control of groups and individuals working at the 'coal-face'.
- Opportunities for learning.
- For personal development and teamwork.
- For making changes to the work environment.

They are bottom-up projects to be encouraged by the quality council and senior managers within the overall framework of the organisation's TQI strategy.

Having initiated a culture of organisational change in which awareness and understanding will have been raised through an increasing number of 'hands-on' improvement projects, the next step is to harness the emerging commitment and gains to improving the value-add capability of the process network itself. This requires developing a quality system to hold and manage the gains and a systematic approach to process improvement.

## Quality system

The implementation of the TQI strategy will also require the early creation of a standardisation infrastructure, in other words a company-wide quality system in which the human, administrative and technical factors affecting quality are defined, clearly understood and managed. In order to achieve consistency these will probably take the form of procedures linked to performance standards. The purpose of a quality system is to establish a framework of reference points, formally recognised throughout the organisation by both external customers and suppliers, to ensure that every time a process is performed the same information, methods, skills and controls are used in the same way. Registration to an internationally recognised quality management standard such as the ISO 9000 system series[8] is an essential building block in any TQI strategy.

The ISO 9000 series defines what is required of a quality system. The standard seeks to identify basic organisational disciplines and specifies the procedures and criteria an organisation must meet to ensure products and services are capable of attaining customer requirements (this can apply equally well to internal customers and suppliers). It comprises five basic components:

1. *Management commitment* – supported by a quality policy which must be understood by all staff.

2. *Organising to manage customer requirements and expectations* – by establishing, in effect, points of interaction between customers and suppliers (external and internal) to provide continuous feedback on customers' requirements and suppliers' performance.

3. *Documenting procedures* – to assure customer requirements are consistently achieved. This documentation should cover how processes are performed, fault handling and queries, change control, inspections and reviews, purchasing arrangements, training, and the handling of exceptions to normal procedures. Documentation does not have to consist of written procedures, flow-charts are just as acceptable and might arguably be more effective in representing the product, service, process and information flows.

4. *Auditing of procedures* – both to confirm that the quality system is operating as prescribed and to identify opportunities for improvement.

5. *Implementing formal corrective actions* – to realign the quality system when necessary and to prevent problems reoccurring.

**Implementing a quality system**

Based on our experience, and the findings of research into the implementation of the BS 5750/ISO 9000 series carried out by Long et al,[9] we suggest the following guidelines for implementing a quality system:

1. *An organisation should be clear on the reasons for the implementation.*
   As we stated in Chapter 1, the quality system has two functions:

   – to underpin the integrity of the company's quality assurance responsibility for providing customers with products and services that consistently meet an agreed standard within the constraints of existing process capability.
   – to provide a reference framework on which to build systematic process improvement activities.

ISO 9000 series registration should not be sought just to satisfy the contractual requirement of a major customer or for marketing purposes. Indeed when most competitors have ISO 9000 certification there is little marketing advantage.

2. *A steering committee should be established comprising all members of the senior management team.* Participation and commitment from all senior managers is essential in order to gain cross-functional support for the project, and to help ensure the smooth implementation and subsequent maintenance of the quality system. Co-ordinating the system's design and implementation is a key task for the quality council. This is achieved by the reporting of the ISO 9000 steering committee into the council.

3. *The design and development of a quality system should be organised as a project.* In common with a quality improvement project the task is better managed by a team. The project needs clear definition, measurable objectives (including registration to the ISO 9000 series), and demands a structured approach to planning and implementation. To assist the steering committee, a project action plan should be prepared at the earliest opportunity. Ideally, this should take the form of a flow-chart of main activities as illustrated in Figure 9.5.

4. *Establish organisation and responsibility structures.* A review of organisational structure and identification of interfaces between functions is essential to the development of a quality system. The findings of the organisation's strengths and weaknesses diagnosis described earlier should provide most of the necessary information. If not, an organisation and activity review must be conducted before system design and implementation work can proceed. This is best done by a qualified quality system auditor against the relevant parts of the ISO 9000 series of standards.

   It will be necessary to prepare organisation charts for all functions and for the organisation as a whole. It is important for all the organisation's members to see how their position relates to the organisation as a whole, and how their activities interface with those of others. Together with process flow charts, these organisation charts will be used to identify the prime activity nodes and information flows of the organisational networks.

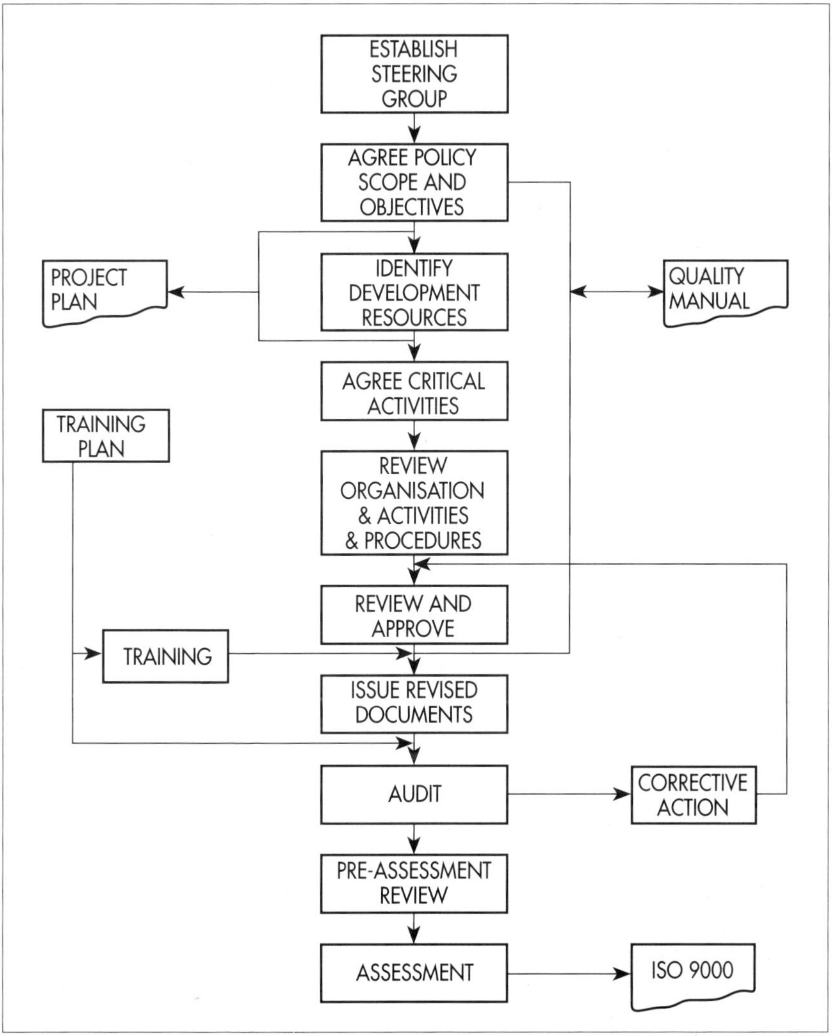


**Figure 9.5** Flow-chart of quality system implementation

5. *Wherever possible, use process flow-charts.*
   There are a number of variants of the classical process flow-chart, including those tailored to an individual company's use which have different symbols to reflect the situation under study. What is important is not the format of the chart and/or flow diagram,

but that the process has been mapped out and is understood by those directly involved in, and responsible for, initiating improvements. Analysing the data collected on a flow-chart can help to uncover irregularities and potential problem points. It is also a useful method of dealing with customer complaints, as it establishes the cause of the break in the customer-supplier chain and rectifies it by means of corrective action. It is also found that in some organisations people are only aware of their own particular aspect of a process. Process mapping helps to facilitate a greater understanding of the whole process, and is essential to the development of the internal customer-supplier concept.

For example, a specific kind of process mapping, termed Q-MAP, has proved to be of particular value to companies such as Garrett Automotive at their Skelmersdale turbocharger manufacturing division. An example of a Q-MAP chart used by Garrett is illustrated by Figure 9.6. The method, which was developed at Garrett by Crossfield from the IDEF (ICAM definition method) methodology, is described in detail by Crossfield and Dale.[10]

6. *Identify and agree the critical activities which are a priority for the quality system.* Critical activities are those activities that are of such importance that they necessitate a systematic approach, and for which documented and controlled procedures must be developed. These will be the prime activity nodes, information flows and interfaces. Critical activities will form the skeleton of the quality system.

7. *Accurate procedures, including operating and work instructions, are required.* These procedures must be practical and easily implemented. Wherever possible, they should document what employees are currently doing. Only where the standard suggests modification should these be introduced. The personnel who are given responsibility for drafting the procedures must be familiar with the requirements of the ISO 9000 series and fully understand the activities to be covered by the procedures.

8. *Implementation requires the co-operation of everyone concerned.* In order to obtain this co-operation, all employees must understand the reasons for implementing a quality system and the benefits to be derived from it. Awareness will be enhanced if the initial organisation and activity review is well explained and sympathetically carried out.

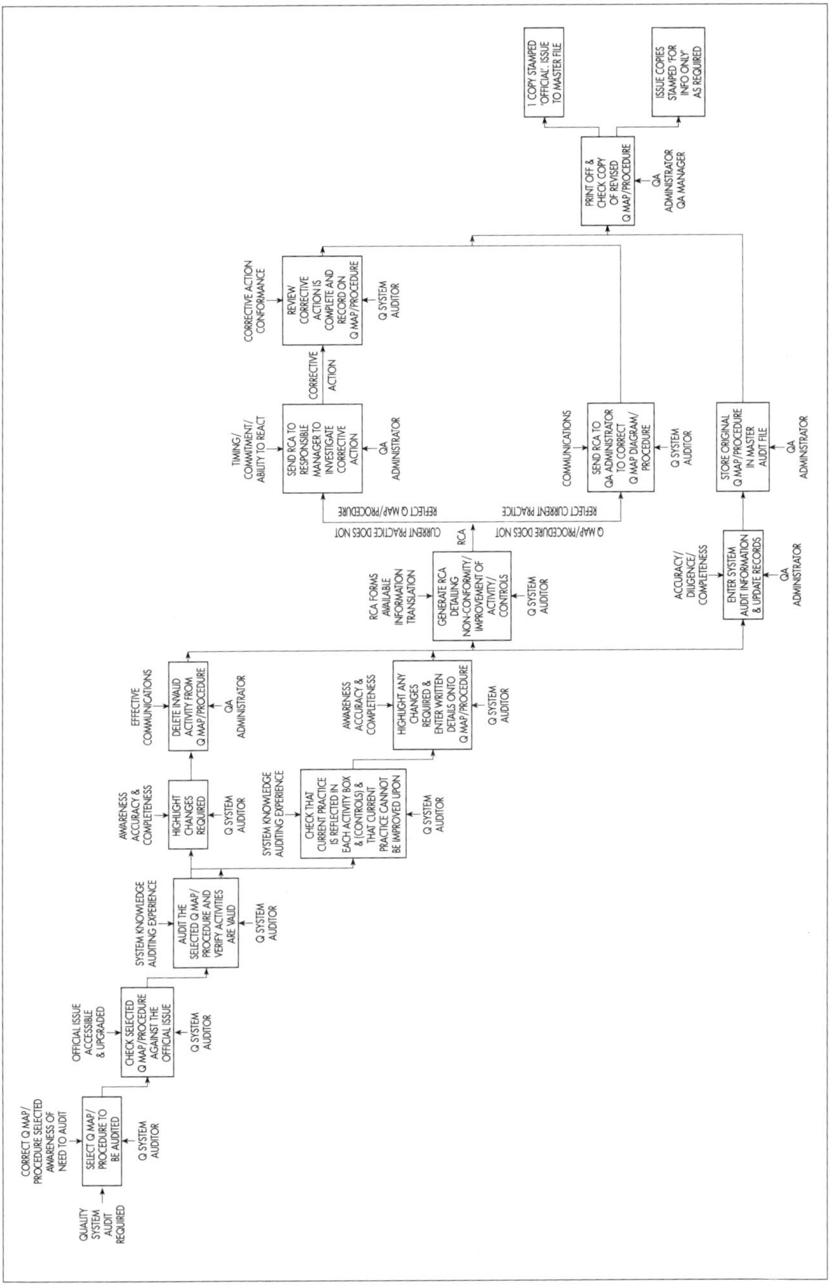


**Figure 9.6** Q-MAP quality system audit

## Auditing quality systems

Formal internal audits of the system, procedures and information flows are necessary to ensure that the quality system remains effective and in line with the strategic direction of the organisation. Traditionally, quality auditing has mainly been perceived as a review of the company's formal quality system. Audits can, in fact, also be applied to products or services, specific processes, and functions.

The auditors should have sufficient respect within the organisation for their findings to be accepted without dispute. They should not normally be responsible to the line management of the area being audited. They should be capable of interpreting the broad requirements of the ISO 9000 series and the company's quality improvement goals, and judging whether they have been applied effectively in the area being audited. Therefore, audit leaders are usually senior and experienced members of the organisation.

Auditing is usually carried out in relation to an agreed set of requirements, e.g., the ISO 9000 series. In the case of a product or service audit, the requirements will be based on customer's need or their perceptions of the product or service. The point is that all audits must have pre-determined requirements. In some cases, these may have to be drawn up to suit a particular audit. Whatever the source, a reference standard is necessary before the audit starts.

During the audit itself, the emphasis should be on gathering factual evidence and on recording specific non-conformances to the requirements. These should be derived by physical examination of the hardware or documentation, and not by discussion only. When an auditor hears a statement of fact, he/she should always be ready to ask for evidence. If possible, non-conformances should be recorded on the spot and signed as accepted by a responsible person in the area being audited.

A checklist should be formulated to assist the auditor. Checklists are not intended to be all-encompassing work instructions for the auditor. They are meant to give each area a consistently uniform audit, and to provide a minimum standard for the auditor. In addition to providing continuity, the checklist notes indicate objective evidence of what was examined. To eliminate ambiguity, all the results will relate to yes/no answers which allow for a quick review during preparation of the audit report.

The audit report should follow a format appropriate to the type of situation. For example, an audit of the quality system would normally be reported to the management of the area concerned in the form of a list of non-conformances, and the actions required to correct the situation. Product or service audits would normally be in the form of defect lists and provide information on the number of defects per product/service. Both these types of audits are carried out on a regular basis. Special *ad hoc* audits are more likely to have a narrative style of report.

### Quality-related information

An effective quality system is more than just a passive process review mechanism: it can be used by the quality council as a dynamic tool for raising performance standards through company-wide dissemination of improvement gains and prevention of problem recurrence. This involves establishing a systematic approach to information management, data analysis, case identification, corrective action, and control and review.

The quality system must be capable of capturing and transmitting data on critical analysis, cause identification and corrective action. Data analysis should set out to reveal precisely and in quantifiable terms:

- What the problem is.
- Where it occurs.
- When it occurs.
- The extent of its effect (in terms of both frequency and cost).

Cause identification is concerned with separating causes and effects using some of the tools and techniques outlined in the next section. Corrective action should not only eliminate the immediate problems being investigated, but should be extended to other products/services which could possess similar inherent problems. The resulting implementation of corrective action, controls and reviews should be put into effect to ensure that actions remain effective. The evolving problem-solving data-bank should be an integral part of the quality system. Information on previous problem-solving activities should be used as part of new design and process reviews. Similarly, quality plans and control plans for new products/services and processes should always take into account any problems previously acted on. The quality system should audit customer feedback (from internal as well as external customers) to ensure that corrective action has eliminated specific problems.

An effective quality system is an interactive framework for systematic information management, standardisation and storing knowledge on which to build further improvement activities. It is a tool for change and should be capable of continuous updating and renewal.

## Process improvement

Process improvement is concerned with improving the capability of individual processes to add value to the overall process stream. Fundamentally, this means reducing and eliminating process waste in all its various forms (see Chapter 1). In statistical terms, waste can be defined as process variation. To manage any process and reduce its variation, the variation must be traced back to its source. In planning process improvement activities the goal is to improve process capability continuously over time, in order to eliminate special and then common causes of variation (see Figure 9.7 from the Ford Motor Company SPC Instruction Guide).[11] By doing so the organisation gets nearer to Taguchi's concept[12] of minimising the loss to society imparted by inefficient transformation processes. Such waste has a compounding effect as it is transferred down the process stream.

In Chapter 1 we argued that any organisation is a network of interrelated processes, both technical and administrative, and that all processes are capable of adding value or generating waste. Therefore, eliminating process variation is just as much concerned with service-related processes, e.g., accounts, marketing, personnel, etc., as with manufacturing-related processes, e.g., cutting, forming, assembly, etc.

Process improvement will involve the use of a selection of quality management tools and techniques, such as SPC, the seven new tools, FMEA, mistake-proofing and department purpose analysis. It is important that management teams are fully aware of the main purpose of each of the tools and techniques they may use. They will need to ask themselves the kind of questions we posed in Chapter 4, recognising that tools and techniques have different roles:

- Data collection.
- Data summary.
- Data presentation.
- Planning.
- Establishing priorities.
- Capability assessment.
- Monitoring and maintaining control, etc.

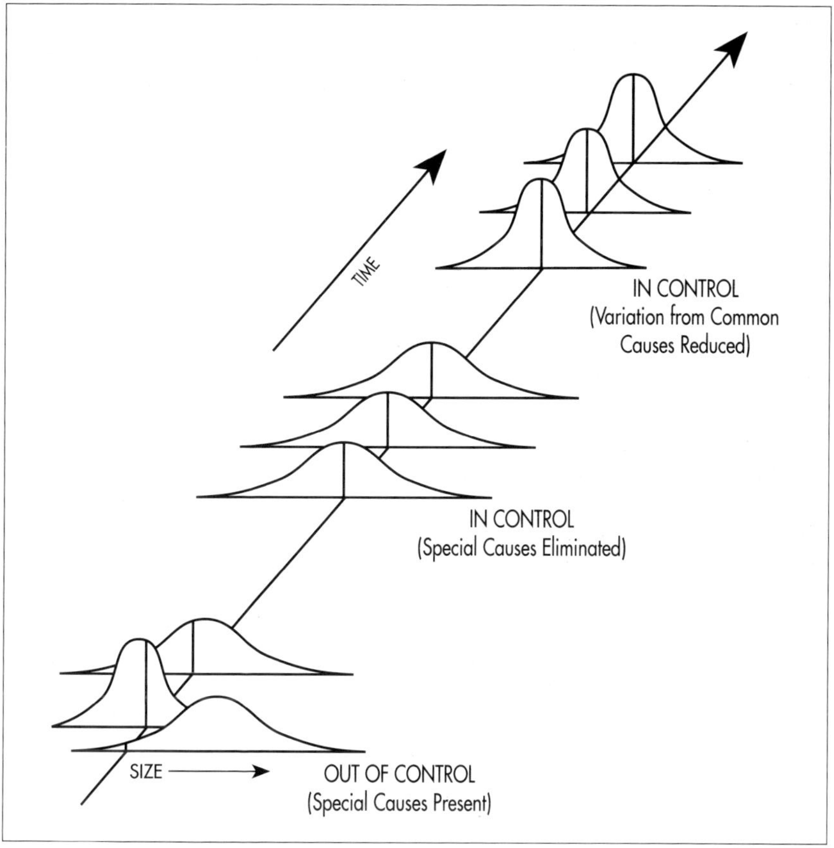


**Figure 9.7** Target process improvement (Source: Ford Motor Company)

A planned approach to deployment is essential.

Process improvement involves three fundamental and complementary tasks:

- Eliminating the special causes of process variation, i.e., *getting the process under control.*
- Reducing and ultimately eliminating the common causes of process variation, i.e., *process engineering.*
- Restructuring and realigning the process management activities in order to improve the entire process stream, i.e., *process alignment.*

**Getting the process under control**

Special causes of variation are any assignable factors which are irregular and unstable, and hence unpredictable. A particular factor may continue to reappear intermittently until positive action is taken to eliminate it. Special causes will usually provide the first challenges to be tackled by an improvement team. Their elimination will involve the use of statistical methods, e.g., SPC, to understand the operating characteristics of the process and the causes of in-process variation, problem-solving tools and implementation of effective countermeasures to prevent reoccurrence.

The use of SPC is essentially a task involving the reduction of in-process variation. The relevant data is usually already available or can be obtained without much difficulty. Information such as scrap, rework, and customer rejects will provide evidence of current problem areas. The seven basic quality tools should be used together to identify the problems which contribute most to customer dissatisfaction (internal as well as external customers) and quality costs. To build a clear picture of how factors which affect the process can contribute to variation, it is vital that everyone involved with the process is given the means to share their experience through process team meetings and brainstorming sessions. Therefore, training in problem analysis and problem-solving skills is a precursor to the use of variation elimination techniques.

Once the process is in a state of statistical control, an assessment of its capability can then be made by comparing the process setting and variation with the specification requirements. Even if a process is operating under statistical control, the amount of variation may be so large that the specification requirements are not met. This means managerial action to redesign the process and possibly the product/ service to achieve sustainable process capability and reliability. The improvement team activity is then applied to eliminating common causes of variation to reduce the variation around the nominal or target value.

SPC is a structured methodology for listening to the 'voice of the process' and using the resulting information to guide process improvement activities. For more details, the reader is referred to books on the subject by Owen,[13] Price[14] and Oakland and Followell.[15]

Statistical methods provide the means for understanding the operating characteristics of a process and the causes of in-process variation, but are only part of the story. Effective counter-measures are necessary to prevent reoccurrence of such causes, which may well involve some form of what might be called 'process engineering'. Improvement techniques such as Design of Experiments, FMEA and 'poka-yoke' are all concerned with process engineering; that is improving process capability through the elimination of all common causes of variation.

**Process engineering**

Common causes refer to the many sources of variation that are always present in varying degrees in different processes. The attention of much of the process improvement activity will be on eliminating common causes, the identification of which may require a deep analysis of the process itself and the objectives of the process owner(s). The solution may involve some restructuring of work activities, process redesign, and resetting of objectives.

Three commonly used process engineering techniques (DOE, FMEA and mistake-proofing) are outlined here. These are not only the only techniques of process engineering, but are probably the most commonly used. Basically, process engineering makes use of techniques concerned with one of three activities:

- Experimenting with process variables to improve process optimisation.
- 'What if' analysis to trace potential causes of failure and their outcomes.
- Making the process less susceptible to failure, particularly through human error.

**Design of Experiments**

Design of Experiments (DOE) is an upstream activity which involves the manipulating of controlled factors at different levels to see their effect on some desired result. The raison d'être for DOE is the proposition that many problems, particularly those involving several variables and interactions, are virtually incapable of solution using traditional experimental methods, changing one factor at a time. DOE methodologies seek to achieve real solutions in relatively short time scales by identifying and controlling the important variable factors, then reducing variation by tighter control or change in design. Two commonly used DOE methodologies are those of Taguchi[12] and Shainin.[16]

The key steps in designing and running such an experiment are:

1. Identify the main parameters/factors, e.g., temperature, the per-

centage of constituents making up a product or mix of material, and interaction. The preparation of a cause and effect diagram may aid this identification. This step should be undertaken by people who are knowledgeable about the process under investigation; this would be done using engineering 'know-how', with some form of instinctive 'gut feeling'.
2. Establish the levels at which each parameter/factor is to be tested.
3. Steps 1 and 2 will identify the orthogonal array to be used (there are a number of popular arrays with perhaps seven covering most applications, and these should be considered as the cookbook). Decide which data are to be placed in each column (see Figure 9.8 from Crossfield and Dale[17]) in relation to the factors being tested, monitoring interactions and detecting noise. It is usual to place those control factors which are more difficult to change in the first and second columns on the left hand of the array. The rows and columns of the orthogonal array are the experimental plan.
4. Organise the experiment and carry it out. This often involves considerable organisation in tracking the products involved it the experiment.
5. Analyse and interpret the results, looking for relevant interactions.
6. Carry out a confirmation run at the optimum settings to validate the conclusions.

Given the appropriate training, these techniques can be used by engineers to achieve significant process improvements, particularly in the manufacture of complex products.

**Failure Mode and Effects Analysis**

FMEA is a systematic quality planning technique for identifying, at the product/service and process design stages, what potentially could go wrong either with a product during its manufacture or end use by the customer, or with the precision of a service. If used correctly, FMEA will assist the establishment of appropriate counter-measures or corrective action which must be applied at the earliest possible stage in the product/service or process development cycle to prevent potential failures.

There are two types of FMEA: design FMEA and process FMEA.

Design FMEA assesses what could, if uncorrected, go wrong with the product in service as a consequence of a weakness in design. The analysis should be carried out in the early stages of a design, before specifications are finalised (after which there could be resistance to

*L8 Orthogonal array*

Standard Orthogonal array

| $L_8$ | 1 | 2 | 3 | 4 | 5 | 6 | 7 |
|---|---|---|---|---|---|---|---|
| 1 | 1 | 1 | 1 | 1 | 1 | 1 | 1 |
| 2 | 1 | 1 | 1 | 2 | 2 | 2 | 2 |
| 3 | 1 | 2 | 2 | 1 | 1 | 2 | 2 |
| 4 | 1 | 2 | 2 | 2 | 2 | 1 | 1 |
| 5 | 2 | 1 | 2 | 1 | 2 | 1 | 2 |
| 6 | 2 | 1 | 2 | 2 | 1 | 2 | 1 |
| 7 | 2 | 2 | 1 | 1 | 2 | 2 | 1 |
| 8 | 2 | 2 | 1 | 2 | 1 | 1 | 2 |

| | Column 1 | 2 | 3 | 4 | 5 | 6 | 7 |
|---|---|---|---|---|---|---|---|
| | U/C | SBTC | 1&2 | Repeat | 1&4 | 2&4 | NSE |
| 1 | with | with | 1 | once | 1 | 1 | 1 |
| 2 | with | with | 1 | 10 times | 2 | 2 | 2 |
| 3 | with | without | 2 | once | 1 | 2 | 2 |
| 4 | with | without | 2 | 10 times | 2 | 1 | 1 |
| 5 | without | with | 2 | once | 2 | 1 | 2 |
| 6 | without | with | 2 | 10 times | 1 | 2 | 1 |
| 7 | without | without | 1 | once | 2 | 2 | 1 |
| 8 | without | without | 1 | 10 times | 1 | 1 | 2 |

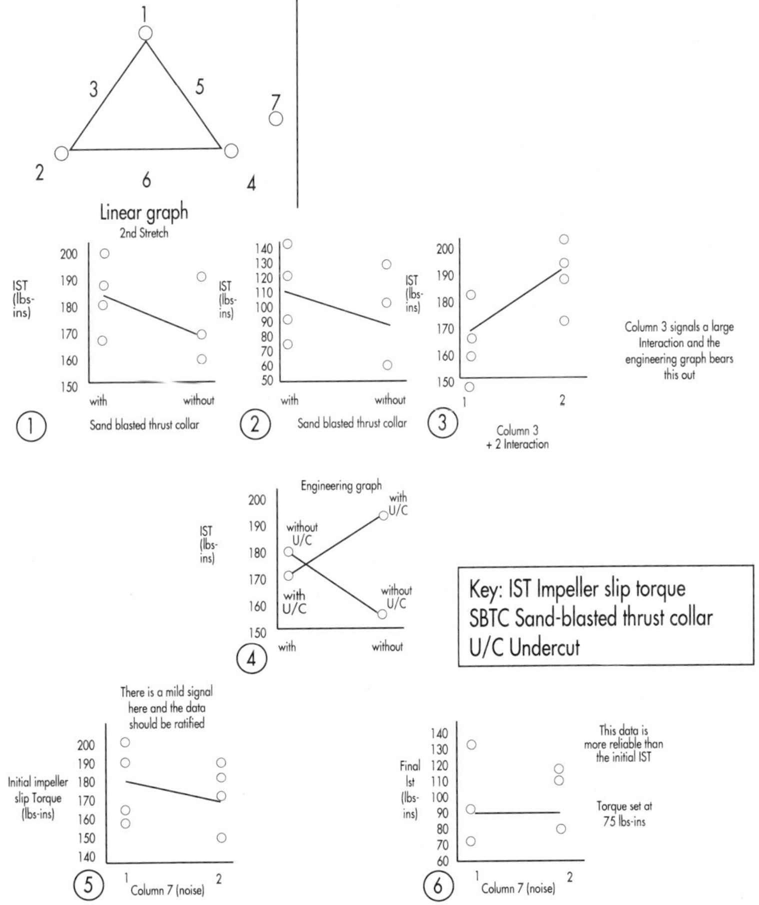


**Figure 9.8** Process activity factors

change in addition to the costs involved).

Process FMEA is concerned with the reasons for potential failure during production resulting from non-compliance with the original design intent or failure to achieve the design specification. The analysis should be carried out before the purchase of tools and equipment, otherwise expensive modifications to the process might be needed later. Process FMEA is a powerful tool for identifying weaknesses in existing processes which have not previously received such analysis. FMEA may be conducted for the whole process or for individual process steps.

With design FMEA, it is assumed that all products/services will be produced correctly to specification (this presupposes that process capability is known). With process FMEA, it is assumed that the product design is correct. These two assumptions make it imperative that both design and process FMEA are carried out by multi-disciplinary teams involving people from the design, production, quality, sales and field service functions, plus any other specialists considered necessary.

**Mistake-Proofing ('Poka-Yoke')**

Mistake-proofing involves finding ways of improving process design to prevent non-conformances. As with FMEA, poka-yoke is a form of 'process engineering' which may involve relatively minor modifications to the process to bring about significant reductions of in-process variation. For a detailed description of Poka-yoke systems and their application the reader is referred to Shingo.[18] Mistake-proofing has two main steps – preventing the occurrence of a defect and detecting the defect. The system is applied at three points in the process:

1. Prevent the start of a process in the event of an error.
2. Prevent a non-conforming product from leaving a process.
3. Prevent a non-conforming part being passed to the next process.

The use of poka-yoke should always be considered when developing new processes, making use of the information developed by FMEA studies. FMEA studies and poka-yoke are complementary variation reduction techniques and should be used together at every opportunity.

FMEA is the ideal opportunity for implementing poka-yoke, but it is not the only occasion when poka-yoke should be considered. Existing processes may be reviewed when there are rejects or downstream customer complaints. Whether it is an individual engineer, project team or quality control circle examining the problem, a knowledge of poka-yoke techniques will enhance their ability to improve process capability.

It is commonly assumed that FMEA and poka-yoke are related solely to manufacturing processes; in fact they are just as applicable to service processes, as indicated by the work carried out under the umbrella of the UMIST TQM Multi-Company Teaching Programme by Catie Gosling[19] at Girobank. Several sophisticated input/output analysis techniques have been developed to assist improvement of service design, such as 'service blueprinting'[20,21] and 'strategy and organisation for service'[22] which feature failure cause and effect data.

**Process alignment**

All processes will have an owner, on whose 'turf' they lie. Unfortunately this otherwise neat arrangement is complicated by the fact that every process is affected by the proceeding process and in turn affects the subsequent process and that these may well be located on someone else's turf. So the constraints to effective process management will probably be caused just as much by inter-process factors, e.g., ineffective product, service, information flows across locations, as by intra-process factors, e.g., poorly defined objectives, invalid performance measures, badly designed work elements, insufficient skills.

This necessitates looking at every process to analyse its purpose, its owner(s), its input and output boundaries, and its impact on other processes. This can be facilitated by using a structured technique such as Department Purpose Analysis (DPA) to:

- Identify staff responsibilities and roles.
- Identify activity purpose.
- Check this alignment against the overall business mission, goals and objectives.
- Examine the internal customer-supplier relationship.
- Identify the major work load elements.
- Expose any elements of waste, i.e., non-value added work.

The idea is to help the process owner and his/her co-workers, firstly to eliminate intra-process constraints by generating a process management checklist of specific objectives, major tasks, major and minor customers, and identified waste improvement opportunities. Secondly, to facilitate a dialogue with process customers to ascertain their view of the process and its impact on their ability to add value.

The process owners will want to know:

- The critical aspects of their service as seen by the customer.
- The customer's likes and dislikes relating to the service.
- Any elements of wasted work identified by the supplier.
- Any performance measures which reflect the critical aspects of the service.

The aim of this exercise is to establish a joint commitment with customers to continuously improve inter-process service and identify any relevant improvement projects. All the improvement opportunities identified are consolidated in a formal process improvement plan agreed by the process owner and his/her staff, and focused on customer satisfaction.

In this way, by encouraging departmental and functional groups to build on developing project management and teamwork skills through formally structuring in-process improvement strategies and extending these across inter-process boundaries, it is possible to set the scene for improving entire streams of processes.

## Process stream improvement

Structured process improvement activities will have initiated cross-process and cross-functional improvement projects. Although a significant step forward, these will be one-to-one activities at the interface of individual processes. So the next step is to align the entire process stream to satisfy and then exceed the requirements of the external customer. This presents a major co-ordination and communications challenge.

The longer the process stream, the greater the diversity of perspectives, values and objectives of individual process owners within the stream. So in spite of the considerable effort required in creating a top-down policy deployment cascade, establishing a quality system reference framework and improving interaction between individual processes, it is unrealistic to expect that process streams with a common set of objectives will automatically follow. This challenge is best met by helping process owners to focus on common targets and align their goals by providing a collective purpose for the process stream based on real customer satisfaction objectives. Quality Function Deployment (QFD) is a powerful tool to help structure process stream purpose.

**Quality Function Deployment**

QFD is a proven methodology for breaking down customer requirements, i.e., the 'whats', into key factors expressed in meaningful business or engineering terms and linking these to realistic process operational requirements, i.e., the 'hows', evaluated against relevant competitive and technical benchmarks. The resulting matrix is commonly called the 'house of quality', for obvious reasons (Figure 9.9). For a detailed description of QFD the reader is referred to Sullivan,[23] Hauser and Clausing,[24] Akao[25] and Eureka and Ryan.[26]

QFD involves two complementary tasks: identifying goals from the customer's viewpoint, and deploying upstream the requirements emerging from these goals. In this way it can be used as a tool to construct chains of 'houses of quality' in which the *hows* for each process in the stream become the *whats* for the proceeding process (Figure 9.10), resulting in a nested set of interrelated process objectives linked to the requirements of the end user. In his article exploring the links between process management and the QFD methodology, Conti[27] visualises QFD as a magnet to align in the

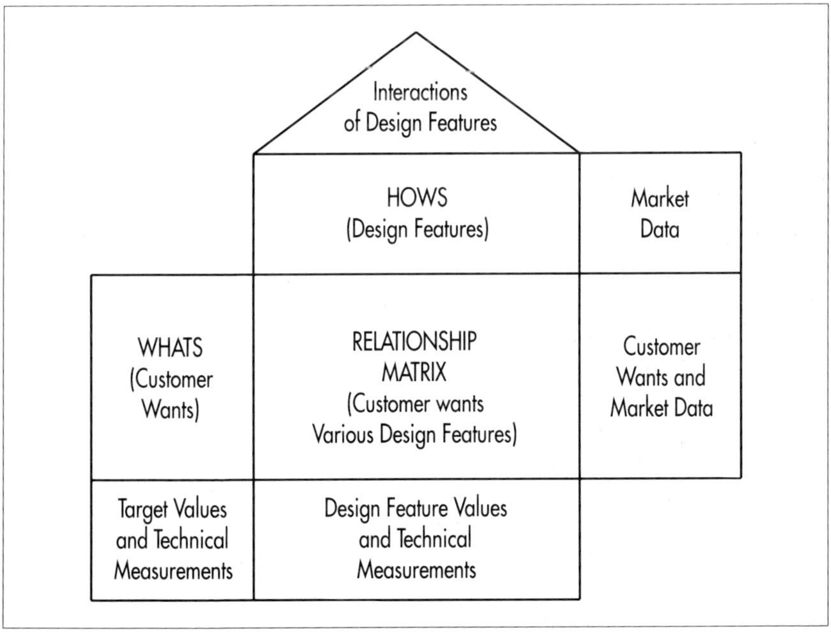


**Figure 9.9** The house of quality

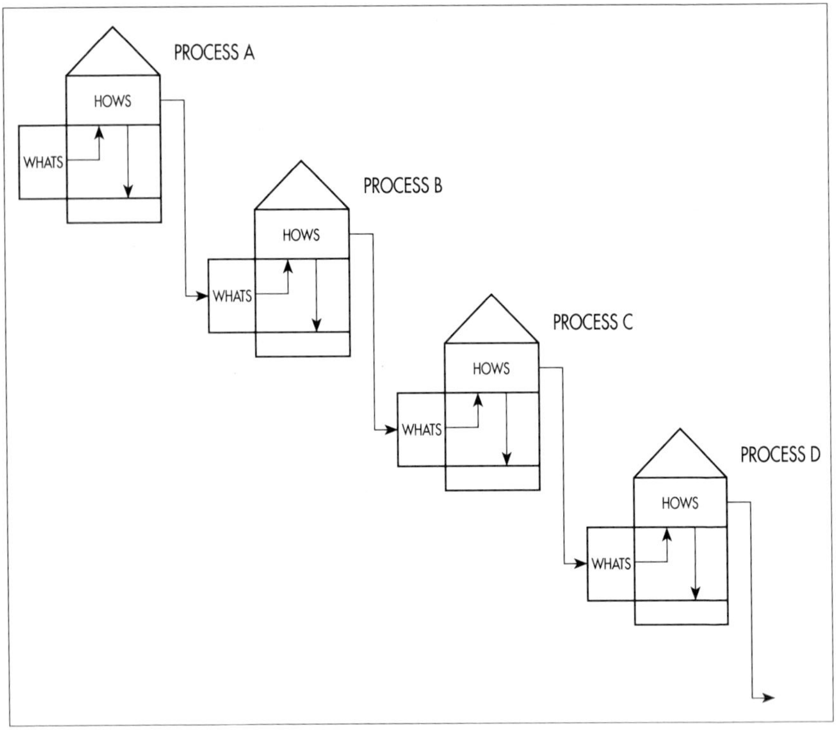


**Figure 9.10** QFD: process stream alignment

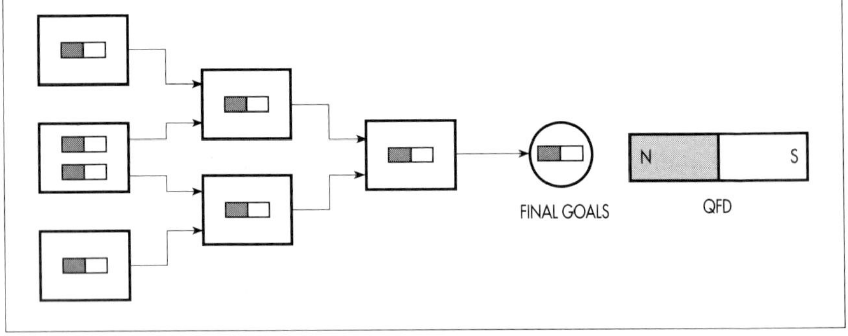


**Figure 9.11** QFD: process stream magnet (Source: Tito Conti, 1989)

same direction all process goals in the stream (Figure 9.11).

If correctly applied, QFD is an effective methodology for aligning process stream objectives and enhancing process stream co-ordination and improvement. However, when planning process stream improvement activity senior management teams need to bear in mind the following considerations:

- QFD is a *tool* for cross-functional management, not a substitute for it.
- The process stream is not confined solely to an organisation itself. External suppliers are also key process owners and are an integral part of the process stream.
- Process stream improvement takes place in a changing, dynamic business environment.

**Cross-functional teamwork**

Teamwork is essential. The application of the QFD methodology is as dependent on teamwork skills as it is on the house of quality charting techniques. If it is to be of lasting benefit, QFD cannot be introduced too early in the TQI strategy. Cross-functional collaboration and integration do not come naturally – trust and experience of the benefits of cross-functional activity have first to be developed. That is why process owners and their co-workers must have satisfactory improvement project teamwork and inter-process problem-solving experience before QFD can be attempted. QFD must be seen as a natural extension of these teamwork activities.

QFD is a relatively simple but detailed methodology which takes time to apply. A major project can take 50-60 hours of meetings involving perhaps up to ten people; time which top management must allow for in its policy deployment rounds. Visible top management commitment is therefore crucial. This means encouraging process owners and their key delegates to participate, keeping the activity a prime management time priority, showing genuine interest in the emerging ideas and opportunities and, where necessary, providing additional resources to exploit these opportunities.

**Co-Makership**

In Chapter 6 we discussed the need to view external suppliers as long-term business partners, involving them in product (and service and process) development, and capitalising on their specialist expertise in a 'co-makership' approach to adding value. Supplier development can be compared with the kind of inter-process improvement activity outlined in the section on process improvement. It starts in the internal purchasing function and extends to the supplier. It involves the free exchange of information, help with internal improvement activities, joint problem-solving and advance quality planning – a sort of DPA in reverse. The next move is really a quantum leap to extend the process stream alignment activity to include external suppliers. This means creating a degree of trust between the parties which goes far beyond the kind of supplier development initiatives at present common in the European business community. But this will be an increasingly necessary trend in an age of rapid competitive change and further technological specialisation.

**Value added chains**

Some distinguished commentators, such as Drucker,[28] have forecast that future business organisations will be based on process streams linked directly to specific customer groups, replacing the traditional function-oriented organisational structure. These will integrate a series of process sets owned by different business companies aligned into streams to satisfy the requirements of a common end-user. These process sets will be both independent (in terms of corporate ownership and technological expertise) and interdependent (in terms of operational integrity and market necessity). A trend very different from the traditional concept of vertical integration – dedicated chains of functional and process hierarchies linked together in an attempt to realise some notion of market *synergy*, but usually rendered obsolete by changing markets and competitive conditions. The kind of process streams envisaged will be lateral, temporary alliances formed to exploit specific market opportunities – which might be called 'value added chains' or to use Kanter's terminology[29] 'PALS' ('pooling, allying and linking' *across* companies).

**Customer delight**

Process stream improvement efforts must be directed at what the customer expects, not what the company thinks the customer expects. Throughout this book we have emphasised customer satisfaction as the *raison d'être* of TQI. We have also stressed the challenge from competitors, and the need to establish competitive advantage by exceeding customer expectations – to delight the customer. But we cannot emphasise strongly enough that in a dynamic competitive environment, customer expectations change. Process stream improvement is a competitive search for adding value and finding customer requirements that have not yet been expressed. This is the basis for updating and regenerating the TQI strategy revolving around goals and deployed objectives which reflect and exceed changing customer expectations.

## Strategic benchmarking

TQI is a never-ending process which necessitates the continuous renewal of goals, objectives and action plans. To remain effective a TQI strategy must, therefore, be dynamic. Benchmarking against best-in-class or industry leaders is an essential part of the strategy renewal process, not only to provide relevant milestones with which to measure progress, but also to generate new ideas and inspire creativity. Benchmarking studies should reveal examples of good practice, and new approaches which can be emulated and turned into effective performance measures. Benchmarking is not a series of one-off studies, but is itself an on-going improvement process which will have an impact on the company's strategic direction.

The whole point of strategic benchmarking is to build competitive advantage by utilising the best available ideas, so it is vital that you know your own company's strengths and weaknesses in order to determine where to start the process. Products, services, processes and practices all affect a company's ability to add value in a competitive business environment. Benchmarking should not be confined to internal comparisons, direct competitors or the company's own industry sector – it must be conducted against best-in-class companies and business functions, regardless of where they exist.

For a thorough description of the benchmarking process the reader is referred to Camp's book on the subject, which contains much advice on how to start and structure the process.[30] Even so, it is worth bearing in mind that there is no one way of carrying out benchmarking studies; an enormous variety of data sources and examples are available. Each study will be different, depending on the internal needs of the company. The terms of reference agreed for a study will help determine the most appropriate methodology, which may involve a combination of data sources. Curiosity and creativity will probably prove to be the most productive tools.

Benchmarking is a comparative analysis of your company's operations and those of industry leaders and direct competitors. You want to know if your company is behind best-in-class operators, by how much and where, and then use this information to close the gap by establishing externally related performance measures to improve knowledge, practices and processes. But it is not enough to quantify the performance gap as a basis for internal measurement – an understanding of how industry leaders actually achieved superior performance is essential to guide new strategic direction.

If managers are to use this information to help guide their improvement activities, care must be taken to integrate benchmark findings into the policy formulation and deployment processes. Benchmark findings will be used to renew the company's improvement goals, operational objectives and the attendant framework of performance measures and review procedures. A database of benchmark findings must be developed and made freely accessible to all levels, so that it becomes an active resource which will add to the organisation's collective continuous improvement experience. This database will have to be constantly updated and renewed in order to reflect the ever-changing external practices and competitive advances. The only way in which strategic benchmarking can become, and continue to be, a dynamic improvement process is for it to be internalised so that it becomes a tool to be used at all levels of the organisation. Specialist advisers can help, but focusing on external practices must become a company-wide responsibility. To help achieve this goal, people close to the 'coal-face' must be encouraged to initiate and sustain their own benchmarking activities. This means encouraging curiosity and creativity, and harnessing and sharing the output for the collective good of the organisation.

Benchmarking is a process of continuous renewal and could, therefore, be described as a strategy of setting new performance standards (comparable to best-in-class) and improving on them. Each advance represents a new milestone of progress – progress along the third dimension of TQI (Figure 9.12). The organisation must benchmark its progress against the six strategic dynamics of TQM:

- Customer care.
- Competition.
- Leadership.
- Suppliers.
- People.
- Money.

**Some benchmark factors**

We list below some of the relevant factors which might be used in developing benchmarks:

- Customer care
  - Causes of satisfaction: what makes the customer come back for more? The causes and factors of customer satisfaction change

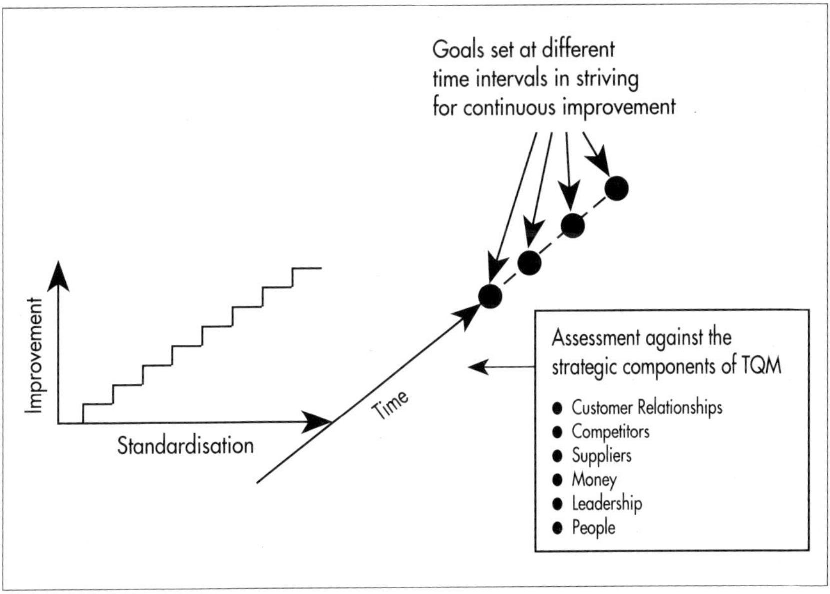


**Figure 9.12** Strategic benchmarks

over time (usually more quickly than commonly realised).

- Causes of dissatisfaction: what factors turn the customer off buying your product or service? (You only have to ask and you will receive answers – lots of them).
- Perception: are you closing the perception gap between the customer's expectations and what you think his/her requirements are?
- Proximity: relationship with the marketplace in a macro sense; closeness of fit with regard to managerial competence and technology, of which ability to satisfy customer requirements is a function.

• Competitors

- Sources of pressure: where are the competitive pressure points and how do they affect your market offering and customer perception of it? How do these change over time?
- Strengths and weaknesses: in relation to your competitive offering (including innovation) and how they change over time.
- Market proximity: their closeness to the market with regard

to competence and technology.
- Market responsiveness: responsiveness to changing customer expectations and technological opportunities.

- Leadership
  - Skills: the skills required for effective leadership and the approaches to leadership.
  - Improvement opportunities: how well improvement opportunities are championed, facilitated and led.
  - Policy deployment: the scope and effectiveness of policy deployment.
  - Diffusion: scale of, and approaches to, extending leadership opportunities throughout the organisation, also the facilitation of cross-functional management.

- Suppliers
  - Key competencies: in terms of technological expertise and market experience, which enable suppliers to add value to your company's ability to delight the end-user.
  - Perception: are you closing the perception gap between your company's expectations and what suppliers think your requirements are? Are your people behaving in a manner outlined in the collaborative agreement?
  - Supplier culture: the progress of your suppliers in developing their own TQI cultures. Are suppliers approaching your company for collaborative partnership agreements?
  - Responsiveness to change: suppliers' ability to cope with constant changes in technology, customer requirements and competitive pressure.
  - Development potential: how many of your suppliers have the potential to develop alongside your company as long-term business partners? How successful is your company at nurturing partnerships, i.e., turning a supplier base into a supply community?

- People
  - Skills base: the development of an underlying skills base throughout the organisation that is aligned with the vision and mission. This means the organisation's success in developing a reservoir of skilled and talented people.
  - Education and training: the effectiveness of the education and training media and the efficiency of the knowledge/skills dif-

  fusion processes.
  - Change agents: the ability of and the opportunities for people in the organisation to manage, change and act as innovators of change.
  - Participation and responsiveness: of people throughout the organisation in TQI activities.
  - Cultural perception: is your company closing the perception gap regarding organisational culture? How close are the vast majority of the organisation's members to owning a collective sense of purpose?

- Money
  - Added value: the areas, opportunities/challenges and scale of increase.
  - Non-added value: the areas, opportunities/challenges and scale of decrease.
  - Measurement tools: for identification and quantification of value parameters; the effectiveness of their application, innovation and development; integration with data base and data integrity improvement activities.

## Summary

TQI is not a one-shot motivation programme. It is a key competitive business strategy and must be treated as such. It must also be integrated with the organisation's other business strategies. This requires careful planning by and the active involvement of the senior management team led by the chief executive.

In this chapter we have outlined a strategic framework for the implementation of TQI involving eight steps:

- Condition setting.
- Policy formulation.
- Policy deployment.
- Improvement projects.
- Quality system.
- Process improvement.
- Process stream improvement.
- Strategic benchmarking.

Planning for TQI will inevitably be time-consuming, but time spent on condition setting and policy formulation will not be wasted. The temptation to rush through the planning phase or reduce it to a bare minimum should be resisted. A clear understanding of what is involved in implementing TQI, together with an unambiguous direction or mission, is essential for success.

This mission will have to centre on two directions: a vertical deployment of policy to ensure that everyone pulls in the same direction, and a horizontal deployment of processes to align all process interfaces and relationships towards enhancing customer delight. Delighted customers, supplier partnerships, transforming leadership and empowered people are the desired outcomes of a TQI mission. Progress towards these must be made both visible and measurable.

Structured incremental improvement, systematic application and standardisation of the gains, and an external focus will have long-term benefits for the organisation. But this does not come in the form of a convenient package that can be bolted-on to the existing organisational structure. To evolve as a world-class competitor it must become a learning organisation, perhaps developing a new organisational form, but certainly creating an open leadership style. This is the basis of the cultural change demanded by Total Quality values. This takes time and patience.

## References

1. Lascelles, D. M., and Dale, B. G., 1988, "A Study of the Quality Management Methods Employed by UK Automotive Suppliers," *Quality and Reliability International*, **4** (4), 301-309.
2. Dale, B. G., and Plunkett, J. J., 1991, *Quality Costing*, Chapman and Hall, London.
3. Campbell, A., Devine, M. and Young, D., 1990, *A Sense of Mission*, Hutchinson Business Books, London.
4. Snijders, L. W. M., 1990, "Total Quality Management, A Never Ending Story," *Proceedings of the Second European Quality Management Forum*, European Foundation for Quality Management, Eindhoven, 87-94.
5. Payne, B. G., 1990, "The Quality Improvement Process in Service Areas," in Dale, B. G., and Plunkett, J. J. (Eds), *Managing Quality*, Philip Allan, Hemel Hempstead.
6. Van't Haaff, A., Unpublished presentation, European Foundation for Quality Management, Eindhoven.
7. Juran, J. M., 1988, *Quality Control Handbook*, McGraw-Hill, New York.
8. ISO 9000, 1987, *Quality System*, International Organisation for Standardisation, Geneva.
9. Long, A. A., Dale, B. G., and Younger, A., 1991, "A Study of BS.5750 Aspirations in Small Companies," *Quality and Reliability Engineering International*, **7** (1), pp. 27-33.
10. Crossfield, R. T., and Dale, B. G., 1990, "Mapping Quality Assurance Systems: A

Methodology," *Quality and Reliability Engineering International*, **6** (3), 167-178.
11. Ford Motor Company, 1985, *SPC Instruction Guide*, Ford Motor Company, Brentwood, Essex.
12. Taguchi, G., 1986, *Introduction to Quality Engineering*, Asian Productivity Organisation, Tokyo.
13. Owen, M., 1989, *SPC and Continuous Improvement*, IFS Publications, Bedford.
14. Price, F., 1984, *Right First Time*, Gower, Aldershot.
15. Oakland, J. S., and Followell, R. F., 1990, *Statistical Process Control*, Second Edition, Heinemann, London.
16. Shainin, D.,
17. Crossfield, R. T., and Dale, B. G., 1991, "Applying Taguchi Methods to the Design Improvement Process of Turbochargers," *Quality Engineering*, **3** (4), 501-516.
18. Shingo, S., 1986, *Zero Quality Control: Source Inspection and the Poka Yoke System*, Productivity Press, Cambridge, MA, USA.
19. Gosling, C., Rowe, S., and Dale, B. G., 1992, "The Use of Quality Management Tools and Techniques in Financial Services: An Examination," *Seventh OMA Conference*, UMIST.
20. Shostack, L., January/February 1984, "Designing Services That Deliver," *Harvard Business Review.*
21. Shostack, L., 1987, "Service Positioning Through Structural Change," *Journal of Marketing*, **51** (1).
22. Dale, A., and Wooler, S., 1988, "Strategy and Organisation for Service: A Process and Content Model," in Brown, S. W., Gummesson, E., Evardsson, B., and Gustavsson, B. O. (Eds.), *Service Quality: Multidisciplinary and Multinational Perspective*, Lexington Books, Lexington, MA, USA.
23. Sullivan, L. P., June 1986, "Quality Function Deployment," *Quality Progress.*
24. Hauser, J. R., and Causing, D., May/June 1988, "The House of Quality," *Harvard Business Review.*
25. Akao, Y., 1990, *Quality Function Deployment: Integrating Customer Requirements into Product Design*, Productivity Press, Cambridge, MA, USA.
26. Eureka, W. E., and Ryan, W. E., 1988, T*he Customer-Driven Company: Managerial Perspectives on QFD*, ASI Press, Michigan.
27. Conti, T., December 1989, "Process Management and Quality Function Deployment," *Quality Progress.*
28. Drucker, P. F., May/June1990, "The Emerging Theory of Manufacturing," *Harvard Business Review.*
29. Kanter, R. M., 1989, *When Giants Learn to Dance*, Simon and Schuster, London.
30. Camp, R. C., 1989, *Competitive Benchmarking: the Search for Best Practices that Lead to Superior Performance*, ASQC Quality Press, Milwaukee, WI, USA.

# Chapter 10
# Becoming the Best

## Introduction

From research work carried out by the UMIST Quality Management Centre during the last 12 or so years, it is clear that the extent to which European companies have adopted and committed themselves to TQM as the ethos of their business is variable in the extreme. It is possible to identify six different levels of TQI adoption (or lack of it). We have termed these levels: Uncommitted, Drifters, Tool Pushers, Improvers, Award Winners and World-Class (Figure 10.1).

These levels are not necessarily the stages through which organisations pass on their TQI journey, rather they are characteristics and behaviour which organisations display in relation to TQI. There are obviously exceptions to these generalised descriptions. Some

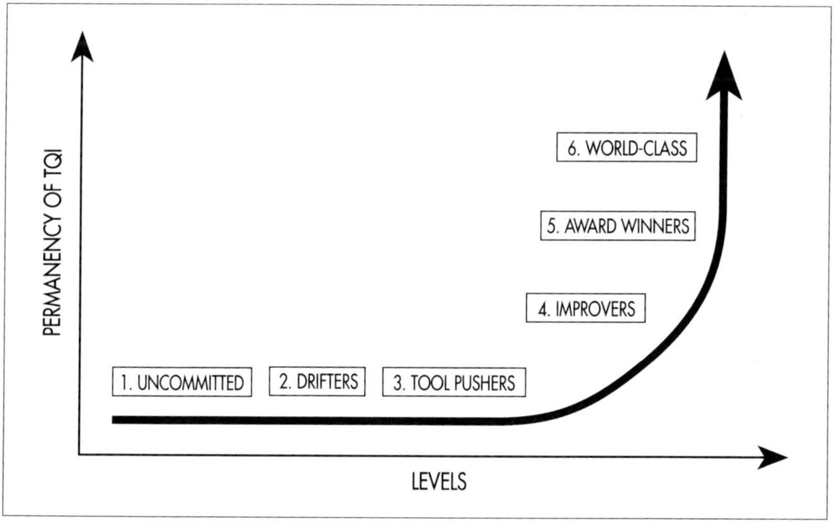


**Figure 10.1** Levels of TQM adoption

organisations which are mid-way between any of the six levels will display the respective hybrid characteristics behaviour.

The six levels are now described.

## Level 1 – Uncommitted

Level 1 type organisations are those which have not yet started a formal process of quality improvement. Their quality management initiatives are usually limited to gaining ISO 9000 quality system series registration and perhaps applying a few quality management tools and techniques, almost certainly as a reaction to customer pressure. Quality improvement is seen both as an externally imposed contractual requirement and as an added cost – a twin threat to be avoided whenever possible. In this type of organisation, quality improvement is not seen as a priority in terms of managerial time and resource allocation. Little, or more probably, no investment in quality management education and training will have taken place. There is also no chance of a leadership culture emerging in the short-term, as senior managers are reluctant to take responsibility for or get involved in improvement activities.

Level 1 organisations are not aware of the benefits of quality improvement. Their management teams (if 'team' is the right word) are usually ignorant of the philosophy and values of total quality and, if they have some knowledge of the concept, are sceptical as to its relevance for them and/or the business.

Level 1 organisations are not necessarily small, immature or unsophisticated. A surprising number of organisations are at Level 1.

They are typified by:

- An overwhelming emphasis on return on sales and net assets employed ('the ratio disease').
- Constant emphasis on meeting output and sales targets.
- Pervading short-termism (as evidenced by frequent changes of priority, lack of investment in people, technology and infrastructure cost-cutting) will be rife, what is sometimes called a 'slash and burn' policy.
- There are likely to be a number of Taylorist type elements, e.g., 'them and us' attitudes, a limited view of 'on-the-job' expertise, inflexible working practices, job demarcation, the potential of individuals is not recognised, embedded in the organisational culture.

- Any quality improvement initiatives which have been taken tend to be 'bottom-up'.
- Their managements and people are usually driven by fear. For example, in a plant of one of the UK's top performing companies (in profit terms), a defective batch of product was hidden from the plant director by the works manager and the production supervisor so that they could dispose of it when he was off-site.

You may say, so what? Do such companies, which are often very profitable, need TQM when they seem to be doing very well without it? They are doing very well – for the time being. But will they still be a major player at the beginning of the next decade? These *uncommitted* companies and their business philosophies are ' dinosaurs' belonging to another economic age – to quote Deming: "They are unlikely to survive the new economic age."[1]

## Level 2 – Drifters

Level 2 type organisations will probably have been engaged in a process of quality improvement for 18 and 36 months and, in general, have followed the available advice and 'received wisdom' on TQM. At this point in the development of total quality the management team will be assessing the progress made. It is also likely that the initial enthusiasm will have worn off and management will be considering ways by which the process may be revived. At this stage, those executives taking a short-term approach will be expressing disappointment that TQM has not lived up to their expectations. Some will even be asking questions such as "What comes after TQM?" In other words, "What do we need to concentrate on next?", "What is the next fad?"

This type of organisation will have probably followed a programme along the lines of Crosby's 14 steps.[2] Only now, having reached Step 14 – 'Do it all again' – they do not know what to do next. In the case of a service, commercial or public-sector organisation they are likely to have started with a customer-care programme. It is usual to find that these initiatives have been rushed and undertaken in a blaze of publicity. At this stage it is not unusual to find such an organisation seeking to employ the philosophy of one of the other quality management experts – a typical comment is: "We started with Crosby and are now viewing the Juran video tapes to see if his philosophy is suitable for our next step forward."

There is a danger that this type of organisation will enter a cycle of programme renewal and decline, moving in ever-decreasing circles of false starts, waning enthusiasm, frustration and disappointment (see Figure 1.10 in Chapter 1).

The characteristics of Level 2 organisations are:

- Quality improvement is still perceived as a programme, not a strategy, not a process.
- There will be no plan for TQM deployment and its development.
- They are too susceptible to outside interventions.
- Management will have undue high expectations of ISO 9000 quality system series registration, and fail to distinguish the difference between meeting the standard and TQM.
- Quality improvement activities will probably be little more than 'off-line' motivational programmes and have made little impression on the company's organisational structure, internal relationships, and overall business direction.
- Any team working will be superficial.
- No real changes in corporate culture will have been made.
- There will still be a reasonably high degree of suspicion and scepticism on the part of both management and staff.
- There will be gaps both in the quality improvement process itself, and in people's understanding of TQM.
- There is likely to be a wide gulf between levels of the organisational hierarchy in their perceptions of TQM, benefits achieved and progress to-date.
- Fear of failure will pervade the organisation.

We call Level 2 organisations *drifters* because they drift from one programme to another almost in stop-start fashion with concepts, ideas and initiatives being reborn and re-launched under different guises. Their management teams frequently try a variety of approaches, often in response to the latest trend, consultancy input and/or what has been picked-up from conference presentations and discussions with other companies. A change of approach can often occur when a senior manager who has been a protagonist of the total quality philosophy and a particular line of thinking leaves the organisation. Their programme conception can appear very creative because their managers are intelligent and articulate people, and some will be genuinely committed to and enthusiastic about TQM. But, so long as they are unable or unwilling to place quality improvement within a strategic business framework it will not be successful.

## Level 3 – Tool Pushers

It is likely that a *tool pusher* will have more operating experience of quality improvement than a *drifter*, perhaps between three to five years.

A level three organisation will usually have ISO 9000 quality system series registration and/or have met the requirements of the quality system standard of one or more major purchasers such as the Ford Motor Company. They will be employing a selection of quality management tools and techniques such as SPC, the seven original quality control tools, quality circles, quality improvement groups, Departmental Purpose Analysis, FMEA and mistake-proofing, and will be in the process of extending their knowledge of some of the more advanced tools such as Design of Experiments, QFD and the seven new quality control tools. An increasing number of organisations at Level 3 are also looking to the criteria of the Malcolm Baldrige National Quality Award or the European Quality Award for Companies to provide an indication to senior management of what is involved in TQM and to give some direction and structure to their process of quality improvement.

Examination of the quality assurance procedures, quality planning systems and the use of quality management tools and techniques will reveal that, in the main, they are being employed and used in a superficial manner, usually in reaction to the demands of powerful customers.

We have also encountered a number of Level 3 organisations which have purchased a tool, a prime example being the Juran tapes, and then followed the recommended advice, i.e., training by module, establishment of problem-solving teams, project by project improvement, etc. However, even though some of these teams have been highly successful, after a period of some 19 to 24 months, the impetus of this type of training has been lost and the Juran training methodology has fallen into disuse. This is not untypical of the way that such companies buy tools, training packages, programmes, etc., and disregard them once the novelty has worn-off, thereby failing to realise the potential afforded by the tool by neglecting to link it into a continuous improvement strategy.

This kind of organisation uses tools, techniques and ideas without thinking through their implications for a quality improvement strategy or its long-term development. (see Chapter 4: Applying Quality Management Tools and Techniques).

The characteristics of this type of organisation are:

- They are forever looking for the latest panacea for a quick-fix (we have seen this happen with quality circles, SPC, Design of Experiments, QFD and, unfortunately, it is now the turn of the Malcolm Baldrige National Quality Award).
- Not all members of the senior management team are likely to be committed to TQM. Those who are will probably not understand its full implications, and there will be considerable variability in their understanding of the subject. Some of these senior managers do not see it as their responsibility to facilitate improvement; there will be a 'what is in it for me' attitude. This will be echoed by middle managers who will typically be saying all the right things, but in their own minds remain unconvinced of the value and strategic importance of TQM, as demonstrated by their day-to-day actions. In improving their area of responsibility, they will give priority to systems and techniques which they consider will have more short-term impact than TQM. This hesitancy and apparent conflict of priorities will be communicated through their actions and comments to first-line supervision and operators. At this organisational level the understanding of TQM and quality improvement is usually patchy and will have a narrow focus.
- The quality effort will be concentrated in the manufacturing/operations departments whilst other departments will remain more or less untouched by the improvement efforts. It is also likely that the quality department will be the main driving force.
- Meeting output targets is the key priority of the majority of managers, and it is likely that there will be conflict between the manufacturing/operations and quality departments.
- The emphasis is on solving current and not future problems.
- There will be repeated claims from some quarters of the organisation that TQM is not working – 'what's new, we did this years ago', and 'the old approach was much better' are typical of the views expressed.

This type of organisation will find it very difficult to sustain the momentum of its improvement initiatives and will be continually on the look-out for new ideas and quick-fixes. There is often a tendency to replace those quality management tools and techniques which have been found to require some real effort and disciplined application to make them work. A Level 3 type of organisation gives the right kind of signals, and presents the requisite image to its customers and suppliers, but under the surface is still a 'fire-fighting' culture not really committed to TQM.

## Level 4 – Improvers

Level 4 type organisations have probably been engaged in a process of quality improvement for between five to eight years and during this time made important advances. They will understand that Total Quality involves long-term cultural change and have recognised the importance of continuous quality improvement. The chief executive and members of the senior management team will have committed themselves to Total Quality through leadership and their own personal actions. They will have formulated their own TQI strategy and have implemented a good deal of it.

As Figure 10.1 shows, there is little difference between *drifters* and *tool pushers* in their adoption of TQI (or lack of it). The significant change occurs with the *improvers*.

Characteristics of this type of organisation include:

- A policy deployment and problem-solving infrastructure will be in place together with a robust and pro-active quality system.
- A high degree of closed-loop error prevention through the control of basic producing and/or service processes will have been attained.
- A long-term education and training programme will be in place.
- Intra- and inter-process improvement activities will be maturing with people looking to improve, on their own initiative, activities within their own sphere of influence.
- The importance of employee involvement has been recognised through a variety of departmental and cross-functional teams and other means.
- Competitive benchmarking studies will have been initiated, and the data used to facilitate improvement activities.

- A leadership culture will be starting to emerge. There will be some strong quality improvement champions, including at least one quality zealot.
- Trust between all levels of the organisational hierarchy will be starting to emerge.
- The preoccupation with numbers will be diminishing.
- The 'hype' which is usually associated with TQM will have faded.

Managers in *drifter* and *tool pusher* organisations often point to the case of those organisations which have made considerable public displays of their commitment to TQM, and yet are turning-in poor financial results. This type of argument will have subsided in the *improver* organisations.

The next step forward involves the management and co-ordination of quality improvement across entire streams of processes – the point at which quality improvement starts to become 'Total'. It is unlikely that the kind of cross-functional culture required to move up to Level 5 will emerge in less than five years; it is more likely to be around 10 years. Process stream improvement and strategic benchmarking activities will not have matured sufficiently. At this stage of its development, TQI may be a focal point, but it will not necessarily have attained *prime* strategic importance, the importance necessary to enable a Level 4 organisation to make the leap to Level 5.

In Level 4 type organisations, TQI will still probably be dependent on a few key individuals to sustain the drive and direction of the improvement strategy. There remains in this type of organisation a danger of lost momentum and failure to 'hold the gains' if: key managers and directors leave; organisational restructuring takes place; and the economic environment and trading conditions become difficult.

Level 4 type organisations are moving in the right direction and are making real progress down the road to TQI. However, they still have a long way to go: as yet, TQI is not internalised throughout the organisation, and the process is not self-sustaining. They are still vulnerable to short-term pressures and unexpected difficulties.

## Level 5 – Award Winners

In their research on the long-term management issues of quality improvement, Williams and Bersch[3] conclude that strong, world-class product and service quality-related competitiveness can only be achieved when an organisation has reached the stage of being able to compete for the Deming Application Prize or Malcolm Baldrige National Quality Award or the European Quality Company Award. This is the point of no return at which quality improvement becomes deep rooted; the organisational culture this will probably take a period of at least ten years to occur in most organisations. The challenge is formidable, and Williams and Bersch estimate in 1989 that probably only 150 or so companies have reached this level of world-class quality. For example, there have been around 180 winners (1992 data) of the Deming Application Prize and the Malcolm Baldrige National Quality Award and the European Quality Award Prize.

Not all organisations which have reached this level will actually have won an internationally recognised company quality award – in fact relatively few have achieved such an honour. But this type of organisation will have reached a point in its Total Quality maturity where it will have developed the kind of culture, values, trust, capabilities, relationship and employee involvement in the business required to win such an award; the point at which quality improvement will have become 'Total' in nature.

Level 5 type organisations will, therefore, have:

- Developed a leadership culture throughout the business that is not dependent on the commitment and drive of a limited number of individuals.
- Successfully implemented change strategies.
- Effective cross-functional management processes and achieved measurable process stream improvements.
- Practice strategic benchmarking at all levels, in conjunction with an integrated system of internal and external performance measurement.
- Made the shift to a more participative organisational culture.
- Managers who have relinquished some of their powers of decision-making to people at lower levels of the organisational hierarchy.
- TQM will be viewed sincerely by all its people as a way of managing the business to satisfy and delight customers (internal and external).

However, although they may appear to form part of an elite, Level 5 type organisations have not necessarily achieved world-class status and made the final TQI breakthrough to the ultimate level of World-Class competitiveness. Rather, the attainment of Level 5 status, that of being seriously able to compete for international recognition, marks the end of an organisation's TQI apprenticeship. Like receiving a black belt in the martial arts, or a scratch handicap in golf, it signifies that the holder has the capability and the potential to make an impact at the highest level – it does not mean that the holder will automatically be a winner. It is a yardstick for measuring progress.

## Level 6 – World-Class

The final breakthrough in TQI is characterised by the total integration of quality improvement and business strategy to delight the customer. Williams and Bersch[3] claimed in 1989 that less than ten companies world-wide, all Japanese, had reached this stage. For example, a company can apply for the Japan Quality Control Medal five years after they have received the Deming Prize, and this award has currently been awarded on just nine occasions (1990 data). Whilst a clear indicator of Total Quality maturity, this award is not the sole qualification for Level 6 status. To stay at the top of the greasy pole of world-class competitiveness takes more than winning the 'second dan' of a corporate black belt.

The handful of organisations which have reached Level 6 and can be considered as *world-class* epitomise the TQI Concept. TQI is concerned with the search for opportunities – opportunities for improving an organisation's ability to totally satisfy the customer. By this stage of Total Quality maturity, which will probably take more than ten years after its initiation, the organisation will be continuously searching to identify more product and/or service factors or characteristics which will increase customer satisfaction through all its networks of process streams. By this stage of the organisation's cultural evolution, the whole focus of its TQI strategy will be on enhancing competitive advantage by enhancing the customer's perception of the company and the attractiveness of the product and service. This constant drive to enhance customer appeal through what the Japanese call *Miryokuteki Hinshitsu* (quality that fascinates) – the almost mystical idea that everything down to the tiniest detail has to be 'just so' – is integral to the concept of continuous improvement. Just like the concept of Total Quality itself, *Miryokuteki Hinshitsu* is a vision, a paradigm, a value framework which will condition an entire organisational culture. This is the breakthrough, the stage at which an organisation finally breaks through to a new mind-set/paradigm: the autonomous and never-ending pursuit of complete customer satisfaction.

It is autonomous in that it has become a personal goal for everyone in the organisation, and an integral part of their everyday working lives. TQI is no longer dependent on top-down, motivational drives and policy deployment, rather it has become a laterally driven network of 'value added chains' or, to use Kanter's terminology,[4] 'PALS' – pooling, allying and linking across organisations.

Never-ending because it has become an obsessive, continuous drive for excellence in all things to delight the customer. TQI is a creative search for opportunities to narrow even further the perception gap between customer and supplier.

At this stage the organisation is very close to the goal of TQI; where the customer's desires and business goals, growth and strategies are inseparable; where Total Quality is the integrative and self-evident organisational truth; where the vision of the entire organisation is aligned to the voice of the customer. In fact, where Total Quality has become the single constant in a dynamic business environment – it is a way of life, a way of doing business.

## The way forward

We live in an exciting era of change and opportunity. TQI is a strategy for change in an environment where the accepted paradigms are subject to constant challenge. It is a strategy concerned with developing an organisational culture in which people are able to meet these challenges and realise the opportunities of change.

With this in mind, you the reader are invited to consider the six levels of TQI adoption and place your organisation at one or another of the levels. You may then wish to consider the following anonymous quotation:

*There are three types of companies:*
*Companies that make things happen*
*Companies that watch things happen*
*and*
*Companies that wonder what happened.*

We wish you luck in plotting your route-map to achieving world-class status.

## References

1. Deming, W. E., 1982, *Quality, Productivity and Competitive Position*, MIT Press, Cambridge, MA, USA.
2. Crosby, P. B., 1979, *Quality is Free*, McGraw-Hill, New York.
3. Williams, R. and Bersch, B., 1989, *Proceedings of the First European Quality Management Forum*, European Foundation for Quality Management, 163-172.
4. Kanter, R. M., 1989, *When Giants Learn to Dance*, Simon and Schuster, London.